In the last few decades, every major pillar of the evolutionary faith has been collapsing all around us. And yet we have evolutionists today saying, "Evolution is a fact. It is not only a fact, it is a most thoroughly proven fact in all of science." Nonsense.

This reminds me of the note in the margin of a preacher's sermon that says, "Argument weak here. Pound pulpit."

That's like standing in the midst of a great building that's collapsed under an earthquake. There's a huge pile of rubble all around, and you say, "Is this not the most significant structure that was ever built?"

But the truth will win out, and I am sure that the day will come when evolution will be recognized to be the pernicious and harmful theory, falsehood, deceit, and fairy tale that it really is. One day people will realize that it is a great and glorious Creator, the almighty, omniscient God, who has created the world and you and me.

That is why I'm pleased to recommend this new book, *Darwinism Under the Microscope: How Recent Scientific Evidence Points to Divine Design* by James P. Gills, M.D. and Thomas E. Woodward, Ph.D. This book provides invaluable evidence that today's leading scientific evidence is Darwin's biggest threat.

Darwin once said that the complexity of the human eye gives him pause about his theory. Well, Dr. Gills has focused his medical studies on the human eye and provides convincing proofs that not only the eye, but also all of Creation, are so complex that the idea they just evolved from chance is impossible. This new book is an important contribution to an important debate.

—D. James Kennedy, Ph.D.
Senior Minister, Coral Ridge Presbyterian Church
Fort Lauderdale, Florida

DARWINISM
under the
microscope

How Recent Scientific Evidence Points to Divine Design

DARWINISM under the microscope

How Recent Scientific Evidence Points to Divine Design

James P. Gills, M.D. & Tom Woodward, Ph.D.

Darwinism Under the Microscope by James P. Gills, M.D. and Thomas E. Woodward, Ph.D.
Published by Charisma House
A Strang Company
600 Rinehart Road
Lake Mary, Florida 32746
www.charismahouse.com

Cover design by Pat Theriault
Interior typography by Sallie Traynor

The editors gratefully acknowledge the contribution of Mark Erickson, medical illustrator, JirehDesign.com, in the design and rendering of the illustrations in this book.

Library of Congress Cataloging-in-Publication Data

Gills, James P., 1934-
Darwinism under the microscope / James P. Gills, Thomas E. Woodward, with Michael J. Behe.
p. cm.
ISBN 0-88419-925-8 (trade paper)
1. Natural selection. 2. Evolution (Biology) I. Woodward, Thomas E.
II. Behe, Michael J., 1952- III. Title
QH375 .G56 2002
576.8'2--dc21

2002011523

03 04 05 06 — 8765432
Printed in the United States of America

Acknowledgments

In a book about Creation and design, acknowledgment and appreciation naturally go to the beautiful mind of the Creator Himself. Too often we refuse to see the manifestation of God in the glory of the sixty trillion amazing cells that comprise each one of us and in the miracle of every being integrated into this vast universe.

This book would never have come to fruition if not for R.T. Kendall who, over a glass of tea under an afternoon sun, encouraged me to print my thoughts on the wonders of life, the travesty of evolutionary theories, and our spiritual blindness to God's majesty in nature.

Richard Swenson, medical colleague and futurist, has written a book, *More Than Meets the Eye*, that directly inspired our own.[1] His vision exceeds the myopia that is restricted to present tenses while it reaches into the intricacies of Creation, leading us to ask, "Who is this Creator?"

Contents

Part Two
Irreducible Complexity and the Detection of Design

Prologue

By R.T. Kendall, M.A., D. Phil.

I will never forget a sermon Dr. Martyn Lloyd-Jones preached at my old church near Oxford when I was a research student there. His text was, "What is man, that thou art mindful of him?" (Ps. 8:4, KJV). Speaking as both a minister and a physician, he proceeded to describe the eye. He demonstrated the utter incredulity of the eye being something that happened by chance or accident and that the only rational explanation for the eye was the Creator God.

When I was given the privilege of reading this book, I anticipated an even more elaborate illustration of the eye, knowing it would be described by one of the most eminent ophthalmologists of our time. What the reader will witness is astounding. You will almost certainly conclude that no unbiased, thinking seeker of truth would any longer hold to Darwinian theories of evolution or their supposed successors.

This timely book is largely a collation of

monographs written by learned men who carefully bridge the gap between the arts and the sciences. Dr. James Gills and Dr. Thomas Woodward are experts in their respective fields. They have combined their knowledge, along with other brilliant minds, in order to produce an easy-to-read volume for any person who is transparently honest in his or her desire to know the difference between fact and fiction, truth and error, what is objective and what is speculation.

What this book further reveals, which may come as a surprise to many laymen, is not only how fragile the theory of evolution is, but also how insecure the scientists are who uphold such. The best of scientists are men at best, but they would prefer to camouflage their biases and fear of being questioned too far. Many of them have simply chosen to believe in evolution. For some it borders on being almost a religion! They are believers in the sense that they have faith that evolution surely must be true. As the reader will see, Darwinism is grounded in *philosophical preference*, not *scientific inference.*

And yet this book also shows that the idea of Intelligent Design of the universe generally, and the human person particularly, is based on physical, empirical evidence rather than a religious idea. In this connection we are introduced to Dr. Michael J. Behe, one of the unsung heroes of our time. His careful analysis of the facts will possibly threaten some and thrill others—especially those who have hoped in the premise that one could believe in a Creator God without being an obscurantist.

On the other hand, no one should suppose that any treatment that exposes the holes in evolutionary theory will be sufficiently convincing that the whole world will suddenly start going to church! When the writer of the Epistle to the Hebrews said, "Through faith we understand that the worlds were framed by the word of God, so that things which are seen were not made of things which do appear" (Heb. 11:3, KJV), he demonstrated an eternal principle. In other words, it is God's decree that we believe His Word by faith. Never will it be possible for one to come to true faith via science, and never will we be able to say, "By science we now believe that the worlds were framed by the word of God"—even if the entire scientific community capitulates to Genesis 1–3. We will never outgrow the need for the Holy Spirit to do His own special, effectual work in bringing men and women to true faith.

This points to the weakness of apologetics. People are seldom converted by apologetics. Apologetics inspires those who already believe the Bible. With few exceptions, apologetics at best only stops the mouths of the opponents to Scripture. But if the reader happens to be fully open to the presentation of true truth, this volume will change minds and change lives.

Foreword

By Richard A. Swenson, M.D.
Physician-researcher and author of *More Than Meets the Eye*

Darwinism is besieged by problems—scientific, statistical, and logical. In an ironic reverse-Galileo effect, evolutionary naturalists often will not permit open discussion of these credibility problems, instead preferring swagger to shut down opposition voices. Using the intimidation factor of pseudo-academia, they "win" their arguments, not by persuasion and good science, but by stifling dissent.

Darwinism Under the Microscope, by a talented group of authors and contributors, opens up the debate. Shining a needed spotlight on Darwinian deficiencies, at the same time this book offers strong credible scientific reasons for faith. Is humanity merely the random product of time + matter + chance? The evidence is stacked mightily against such a nihilistic conclusion. "In the beginning God created..." is not only good theology, but good science as well.

The notion that science and faith are enemies is a myth. Quite the opposite: Science is a friend of faith. There are no scientific facts that obstruct faith. Belief in God does not require a scientific inferiority complex. Coming to faith does not require leaving the intellect behind. Science augments faith, revealing staggering levels of sophistication in order, complexity, and design. The scientific reasons for faith today are prolific—we might even say, in an inflationary cycle.

Let the record show: God is impressive. A spectacular God will give rise to a spectacular Creation. At the end of every scientific inquiry we find the fingerprints of a creative Deity. To see with wonder is one of the great joys of life. Read. See. And rejoice.

Introduction

By James P. Gills, M.D. and Thomas E. Woodward, Ph.D.

Does Darwinism deserve any longer the status of "established scientific fact"? Is Darwin's "tree of life" real or mythical? Most important of all: Was the cell's amazing complexity crafted by genius or by genetic lottery?

The scientific controversy over these questions first exploded into the gaze of the public in 1996 with a "cultural earthquake"—the publication of *Darwin's Black Box* by Lehigh University biologist Michael Behe.[1] Professor Behe argued that, using Charles Darwin's own proposed test, new molecular evidence, which revealed an unanticipated "irreducible" type of cellular complexity, had propelled Darwinism in a fatal breakdown. Nearly a decade has passed since then, and Behe's scientific heresy has now penetrated widely into the American consciousness. Suddenly, at the dawn of the twenty-first century, this controversy has begun to experience a rapid and momentous surge of growth and influence.

Even though the jolt of Behe's critique has been absorbed into the Darwinian consciousness and scientific teacups have been restored to their proper places, the sting of his radical diagnosis has not faded away in the intervening years. Moreover, Behe's work, alongside that of Phillip Johnson and William Dembski, is now recognized as just the leading edge of an entire "wedge" of several hundred scientists, philosophers, engineers, historians, and other academicians who constitute a potent new movement in science. This revolutionary movement has become known as "Intelligent Design."

Awareness of this movement and its ideas has grown rapidly at both the popular and the academic levels of society. The "paper of record," the *New York Times*, has covered the rise of Intelligent Design step by step. After a rather positive review of Behe's book, the *Times* twice invited the author to write his own opinion piece on the detection of design in biology. A turning point was April 6, 2001, when *Times* science writer Nicholas Wade penned an important front-page story that distinguished Intelligent Design from the older "Creation Science" movement that had received so much publicity in recent decades.[2] Wade's article was remarkably evenhanded and pointed out the key role of William Dembski of Baylor University, who labors as the chief mathematical theorist of Intelligent Design. In the 1990s, Dembski forged a unifying methodology for the movement when he set forth his "explanatory filter"—a rigorous system for detecting intelligent action in any phenomenon in the universe.[3]

More recently, the American Museum of Natural History took official note of the movement when it staged a dual airing of design ideas, through a dedicated issue of its magazine (in April 2002, with three design articles and four rebuttals) and through a sold-out public debate at the museum that same month, in which both Professors Behe and Dembski participated. In addition, several states have explored the educational implications of design. Alabama now pastes a "disclaimer" in its biology texts. (You can read this disclaimer in Appendix B.) Kansas earned the wrath of newspaper editorialists and scientists (for a while) by drawing up guidelines that made large-scale evolution (called "macroevolution") *optional* for local school boards to require in their curriculum.[4] As this book goes to press, the state board of education of Ohio is considering a change in their high school science standards, which would strongly encourage teachers to present the "Darwinism-vs.-Design" controversy in high school biology classes of Ohio.

When state boards of education wrestle over the "Darwin controversy," when the leading fossil museum in America airs critiques emanating from design theorists, and when the *New York Times* gives respectful coverage to the scientific skepticism of the "wedge," then it can be said: "We're not in Kansas anymore." In view of this profound and historic transformation, it is urgent now—indeed, it is a matter of civic duty—that the public gain a better grasp of the evolutionary controversy. What is needed, simply, is to place *Darwinism under the microscope.*

To name this book, we chose the phrase italicized in the sentence above, recognizing that "Darwinism under the microscope" can (and should) be taken two ways. One way, already alluded to, is that Darwinism can no longer be granted a relaxed acceptance as it was for nearly all of the twentieth century. A new century has opened with nearly every aspect of Darwinism, every one of its branches, every line of its thought, under an intense scrutiny never before experienced. This is a healthy development, and long overdue, because an unexamined science tends to decay into a lazy, unaccountable caricature of science—one that substitutes suppositions and speculations for rigorous explanation.

The second sense of our title is that, with the advent of electron microscopy and molecular biology, Darwinism is now subject to a more penetrating "microscope" level of testing than ever before. Scientists can take the public on a tour of the exciting world of "nano-machines" within the cell, now that their function has finally been elucidated. One of the themes of many chapters in this book is that the myriad details that have come to light though this microscopic gaze have not only led to an erosion of confidence in Darwinian explanation, but *more significantly, they have bolstered the inference to Intelligent Design—the notion that life was designed by a prodigious intellect.*

The book is divided into two sections. The first half, "The Unraveling of Darwinism," provides an introduction and traces the rise of the entire controversy in the 1990s and beyond. This is accomplished first through a pair of introductory chapters—a lively dialogue in a university cafeteria titled "Coffee, Darwin, and Design," followed by Mark Hartwig's elegant overview chapter, "Challenging Darwin's Myths." (In tandem with Dr. Hartwig's article, we suggest that you read "Science vs. Science," Lynn Vincent's dazzlingly vivid profile of the top leaders of the Design Movement. This piece, sometimes hilarious but always delightful and penetrating, is located in Appendix A.) Next, James P. Gills, in his chapter "The Magnificently Complex Cell," details the stunning complexity and beauty found in a variety of cells—all of which constitute compelling evidence for design.

Next comes a pair of concise manifestos: First, Charles Thaxton's "DNA, Design, and the Origin of Life" is a classic and timeless presentation of the argument for design based on the informational nature of DNA. Second, Phillip Johnson's "Darwinism on Trial," a glittering scientific gem never before published, is one of the most powerful summaries ever penned showing the evidentiary crisis of evolution and the urgent need of an overhaul of the dogmatic teaching of Darwinism. Part One is rounded out by Thomas Woodward's overview of the problem of transitions and the Cambrian Explosion: "Of Canadian Oddballs and Chinese Monsters."

Part Two, titled "Irreducible Complexity and the Detection of Design," shifts from the unraveling of Darwinian explanations to the positive evidence that biological complexity was indeed designed. Woodward's opening chapter, "Meeting Darwin's Wager," provides an overview for this section, focusing on the chain of events and discoveries that led Michael Behe to research the breakdown of Darwinism and to create a new proposal for detecting design. Behe's own chapter, "The Modern Intelligent Design Hypothesis: Breaking Rules," summarizes his proposal and responds to a range of criticism that he has received since his book was published. George Ayoub's chapter, "On the Design of the Vertebrate Retina," responds to the specific charge (leveled by many Darwinists) that the "backward wiring" of the rod and cone cells betrays a clumsy, patched-together origin by mutation and selection rather than by a wise designer. Ayoub shows why the vertebrate design is, in fact, an example of superior wisdom and engineering.

Next, Thomas Woodward's report, "The Mystery of the Yeast Genome," shows why the discoveries of geneticists who unravel the secrets of entire genomes often wind up undermining, rather than supporting, the Darwinian assumptions so prevalent in biology today. In the penultimate chapter, design theoretician William Dembski explains why the time has come for science to open up a new category of causation—design—in order to have any effective explanation for a unique type of phenomena called "specified complexity." He also details a rigorous method by which design can be reliably detected. Last, in his conclusion, James Gills draws together the many threads of the book and helps the reader sort out the powerful implications for personal reorientation that flow from the compelling case for Intelligent Design.

Each of these chapters has its own unique character and personality. Some are a bit more challenging than others, but all are packed with powerful insights and crucial information. If you find yourself taxed a bit by some "scientific jargon" here and there, we suggest you just switch to the "skim mode," or just skip to the next section (or chapter). Consider this book your own delicious smorgasbord that you can move around in.

It is the strong hope of both the editors and the authors of individual chapters that *Darwinism Under the Microscope* will prove to be enlightening, stimulating, and thoroughly rewarding, both intellectually and spiritually. Darwinism, like a long-locked attic, is badly in need of airing out. We hope that fresh breezes of scientific discovery, bold inference, and radical reassessment will blow vigorously through an area of science and culture that bears the musty odor of stale slogans rooted in ignorance and of dogmas long shielded from the light of open inspection. Both the scientific enterprise—and the human spirit—cannot fail to flourish as we pry open the window and peer into the waiting microscope.

Part One

The Unraveling of Darwinism

Teach Darwin's elegant theory. But also discuss where it has real problems accounting for the data, where data are severely limited, where scientists are engaged in wishful thinking, and where alternative, even heretical, explanations are possible.

—MICHAEL BEHE, IN "TEACH EVOLUTION—AND ASK HARD QUESTIONS,"
THE NEW YORK TIMES, AUGUST 1999

Coffee, Darwin and Design

By James P. Gills, M.D. and Thomas E. Woodward, Ph.D.

A midafternoon gust of frigid wind sent the limbs of a large oak gesturing wildly, releasing a fresh cluster of bright red leaves from the half-stripped tree. Soaring briefly on their maiden voyage, the leaves fluttered downward to join the crimson carpet below the tree.

Just fifty feet from the tree, on the other side of an expansive glass wall, sat Jackie, a freckle-covered, petite grad student, hunched over a steaming cup of coffee from the cafeteria's self-serve bar. She was dressed in faded jeans, sneakers, and a thick blue sweatshirt that matched the late afternoon sky framing the oak tree.

Just minutes earlier she had jogged through the windswept yard that sprawled downhill between the biology building—affectionately dubbed "Bio" by students—and the nearly deserted university cafeteria. The wind had left her slightly disheveled, creatively twirling and looping the bright red hair that flowed straight down her back and shoulders.

A quizzical look covered Jackie's face as she rifled through every corner of her backpack. Finally, she smiled as she found what she wanted—a lecture outline titled "Irreducible Complexity: Evidence of Design?" that she had snatched from a seat at the back of the main lecture hall at Bio as she left for the cafeteria.

As Jackie finished stirring her coffee, she scanned the main points and studied the diagrams. Slowly she took her first sip from the cup, then pulled it back suddenly and grimaced slightly. She looked at her watch—Kiera should arrive at any moment. Glancing behind her, she smiled at her lanky, black-haired cousin who was striding toward the table, carrying her own steaming cup of coffee.

Kiera set down the cup and slammed down her mountain of books. "Can you believe that garbage?"

"I know, the coffee's terrible," Jackie smiled.

"No, I mean our guest lecturer."

"I thought Dr. Behe gave a good talk."

"If you're sitting in a pew," Kiera grumbled as she pulled out a chair to sit down. She peeled off her burgundy, nylon fall jacket and laid it on the back of an adjacent chair. Her black hair was pulled back into a ponytail, and she wore a mint-green sweater over an ivory blouse, with black slacks that matched her hair.

Jackie furrowed her brow, and replied, "I'm sorry you didn't care for his presentation."

"No offense, Jackie, but when will you creationists get it?"

"Get what?"

"Get a scientific grip on physical reality. Like irrefutable proof that Darwin got it right the first time."

Peppered Moth—Proof?

Kiera paused to fumble through her stack of books until she found the one she was looking for. She flipped to a page with a picture of the famed peppered moths of England.

"What are you so desperately looking for?"

"This!" Kiera pointed to the picture of the light and dark moths perched on a tree trunk whose grooved surface was mottled with pale green lichen. "Although it's something you probably don't want to face."

"I always said that you were a bit slow."

"What do you mean? How can you deny the evolution that has been observed in the wild?"

Jackie paused and took a deep breath. "I don't want to sound arrogant, but you really do need to catch up."

"On what?"

"On your reading! Look, the peppered moth never *proved evolution anyway.* It was just a back-and-forth adjustment between two varieties in the same species. And both varieties were there all along. Did you notice?" At this point, Jackie slowed to emphasize each of her next words: *"We didn't actually see one arise from the other."* Suddenly, her voice lowered to a hush. "And now, scientists are quietly whispering to each other that there's an even more embarrassing problem."

Kiera blurted out, "And what might that be?"

"The whole story of the peppered moths hinges on the background color of tree trunks, but evolutionists have finally admitted—after knowing it for quite a few years—that this species of moths doesn't even rest on tree trunks. The evidence now shows that when they rest, they hang out in the canopy of the trees, on the underside of the higher twigs and branches."

"How could that be?" Kiera firmly tapped the glossy page she held open. "This is an actual *photo* of two peppered moths. And unless I'm seeing things, *that is a tree trunk!*"

"Sorry, but those photos were staged. Dead or torpid moths were placed on trunks for the photographer. Sounds like fraud to me."

Kiera's face betrayed disbelief. "Who says this?

"Here." Jackie slid a book across the table toward her cousin. "Catch up on the network news with Wells' book *The Icons of Evolution.* He has a whole chapter on the mess of deception and misinformation about peppered moths. It's practically a scandal how they keep publishing this—what was *your word? Garbage*—in every biology textbook."

Kiera pulled the book toward her slowly and eyed the cover suspiciously. "Wells—hmm...never heard of him." Kiera handed the book back to Jackie and added, "He's probably a know-nothing crank."

Jackie tucked the book into her backpack. "Wells doesn't quite qualify as a nobody from Podunkville. He got his Ph.D. in biochemistry a few years ago from a little place called the University of California at Berkeley. Oh, and he already had a Ph.D. from Yale."

Kiera shook her head. "That doesn't mean *anything.* You made a point to tell me about Behe's Ph.D. in biochemistry from an Ivy League school, but his ideas were nothing new. He just wheeled out and dusted off the old watchmaker analogies from Paley in the 1800s. *Pure theology!* You call that science?"

"Help me out here. What *didn't* you like about his talk?" Jackie paused, and after a moment of silence, added, "Try to be specific."

Kiera glanced at the oak tree as another gust sent it into a fresh gyration, strewing leaves in every direction. She gathered her thoughts, staring in the distance, and then spoke as if she were quietly confronting a complex but frustrating enigma. "I don't know... Maybe it was Behe's display of the whole intellectual wasteland that is Creationism, or Intelligent Design, or whatever they call it now."

Jackie paused, pondering how to respond to Kiera's general description. "You seem to have a good grasp on Behe's biochemical explanations."

"Forgive me, Ms. Research Professor wannabe. But we premed

folks still know a thing or two."

"Like ten easy steps to burning a straw man."

Suddenly, Kiera's voice became even quieter but insistently emphatic. "No. Like seeing the inescapable conclusion that evolution is empirically true."

Evidence... What's the Evidence?

Jackie sat up erect in her chair. "Hit me with your best shot. The moths are dead; can you do better?"

"Even I know that as each species develops you can see it passing through the major steps of its evolution."

Jackie's hands shot out sideways in a gesture of disbelief. "What? How so? You saw Behe's closing slide showing Haeckel's famous embryos..."

Kiera interrupted, "Actually, I had to miss the last five minutes, so you'll have to clue me in."

"Well, it boils down to this. Darwin's German bulldog, Ernst Haeckel, *deliberately fudged* the drawings of different embryos to make them look much closer than they were. What's worse, many of the textbooks still print those drawings as proof of evolution, even though the fraud has been known for a century! *And don't say we didn't talk about this already. Remember our chat after Grayson's lecture on embryology?*"

Kiera nodded, "You mean when he showed pictures of human embryos with gill slits?"

"Yes," said Jackie, "but the supposed 'gill slits' in mammal embryos *aren't gill slits at all*—they're called 'pharyngeal pouches,' and they develop into parts of the jaw and inner ear." Jackie shook her head in dramatic disgust. "So much for that stupid mantra—'Ontogeny recapitulates phylogeny'!"

"OK, OK. *But I'm not talking mainly about embryology.* I'm talking about change through time—the patterns that every paleontologist sees in the strata. It's not that hard to understand,

Jackie. *Gradual change occurs over time, through genetic variation, through the vehicle of random mutation.* Those mutations that are beneficial, and those species that have them, survive. Others bite the dust."

Jackie ventured another sip of coffee. "Sounds pretty simple to me. Larry the fast cheetah catches lots of antelope and makes lots of little cheetahs that pass that speed on to other cheetahs. Lenny the slow cheetah gets eaten by Vinny the vulture. *Evolution?* Not! That's *microevolution!* And didn't our speaker talk about the inability of mutation and selection to build molecular machines? Doesn't that implicate other possibilities?"

"Yes, and he represents fringe science. Or should I say, 'pseudoscience.' Grayson doesn't even agree with him."

"Maybe Professor Grayson is open-minded enough to let us hear *all* of the possibilities?"

Kiera suddenly slid out her chair and stood up, posing as if she were standing at a lectern in a science lecture hall. "You know what, audience? They didn't land on the moon. That was just Hollywood magic."

Jackie looked down and shook her head. "Here we go."

Kiera continued with her mock lecture. "And class, not only is the earth the center of the universe, it's also flat!"

"Are you done?"

Kiera smiled as she sat down. "I'm just practicing being open-minded."

Jackie stared at her cousin for a second, then stood up and began to pack her things. Suddenly she stopped, looked at Kiera, and said, "Will you stop *to think* for two seconds?"

"I'm joking; sit down."

Thought Experiments

Jackie slowly sat, then took another gulp of coffee as she thought about how to proceed. She looked Kiera in the eye and said, "Did you ever stop to think that the reason creationists aren't taken seriously by you and Stephen Jay Gould and Richard Dawkins and the rest of the Darwinian inner circle is that the heretics like Behe might be onto something? That a paradigm shift was actually in its early stages, leading to a full-blown scientific revolution?"

"No, because I'm a scientist. And that little picture of a 'paradigm shift' takes us way outside the realm of scientific investigation. Science can only study causes that it can directly observe, *not mysterious intelligences or deities.*"

"Not so."

"Oh? Give me a counter example!"

"Try SETI."

"You mean 'Search for Extraterrestrial Intelligence'?"

"Exactly. If one of our friends at NASA *actually got* a mathematically coded message, like in the movie *Contact*, we'd know there was an intelligent agent out there who sent it without directly observing that agent. The *inference* from observed effect—radio emission—to inferred cause—some sort of intelligence out there—would be solid *without directly observing that intelligence.*"

"Jackie, your SETI analogy is clever, but it stumbles on a simple fatal fact. Any ETs that were discovered by NASA would be more or less 'conventional entities'—fairly familiar biological beings like ourselves, not some mysterious 'force' or 'intelligent agent.'"

"Wait a minute, Kiera. Time out!" Jackie held her hands in the familiar sports symbol of a T. "Am I understanding that you are actually admitting—*finally*—that science can detect—by empirical methods—*an intelligent cause of some phenomenon?*"

"Well, yes, in the special case of SETI, but you can't just leap from there to God!"

"But we're not saying—that is, design theorists aren't saying—that *science alone* can show or prove the statement, 'God made the flagellum.' Yet, at the same time, once the conclusion is clear that something intelligent must have made these complex machines, then the self-enforced isolation of science will just have to crack open. Theologians and philosophers deserve at least an invitation to the table of wider discussion at the point where we conclude that there's 'an intelligent cause' of something in biology."

"My dear Jackie, the theologians have been the *bad guys* in science, not the good guys, and they're the *last people* to invite to a conference on origins. Reread the Galileo story."

"Kiera, you're the one who needs to reread Galileo, since the church was upset with his departure from popular philosophical ideas handed down from Aristotle, not with any opposition to biblical statements. But that's another cafeteria debate. Can I continue with my thought experiment?"

"Go ahead—dream on..."

"What if someone, someone you had told others was a complete 'moron,' wound up discovering something truly revolutionary, something that much of the intellectual world eventually acknowledged as true, even though it went against everything that you had believed in, against everything that you ever conceived?"

Kiera cocked her head to one side, her expression suddenly dead serious. "When you say 'discover,' do you mean discover *evidence*, not just nice little watchmaker theories?"

"No, I mean *a theory that is based in solid scientific evidence*—evidence like the detailed structure of the rotary motor of the flagellum on the back end of a bacterium, or the nuts and bolts of a little hairlike cilium that whips back and forth on the tail of a sperm."

Kiera ventured her first sip of the maligned coffee. "I'm listening..."

"And what if that discovery—of a clearly Intelligent Designer—put some of what you taught as 'absolutely true' to shame, even tore it to pieces?"

Kiera shot back, "Open-minded *zealots* to the rescue," then quickly added, *"OK, Jackie; go ahead."*

"I know you heard the talk. But did you actually listen to Dr. Behe?"

Sudden Leaps and Botched Retinas

"Yes, I listened. He did a so-so job of explaining *irreducible complexity* and replying to the criticism he's gotten. But just because we don't know now how some systems in nature have formed doesn't mean that they didn't form by random mutation over eons of time, and it doesn't mean that we won't figure this stuff out in the next hundred years."

"But Kiera, if there has been all this gradual development over time, why do we never seem to find the transitional critters—especially between the major groups?"

Kiera was not impressed. "That's an old argument—no big deal. See Gould's 'punctuated equilibrium' model. Makes sense to me."

"The problem with that is that we're not just talking about an enzyme here or a protein there. We're talking about things like the human eye and the Cambrian Explosion—sixty or seventy phyla bursting onto the scene out of nowhere."

"And that would be in perfect keeping with Gould's notion that the fossil record indicates long evolutionary equilibrium, punctuated by sudden leaps of progress."

Jackie shook her head in disbelief. "No…We're not talking about *part* of the human eye, we're talking about the *whole* eye, from the protein cascade inside a single rod cell all the way up to a fully integrated, camera-type eye with dozens of cell types and the brain centers to interpret all the input."

Kiera brightened as she saw an opening. "Since you brought it up, I hope that your Designer has a good lawyer."

"Why is that?"

"Because He did a pretty terrible job designing the eye."

"What?"

"It's wired backward!" Kiera grabbed a pen from her bag and started sketching animatedly a layer of rod and cone cells on a paper napkin. "Don't you know that the retina is incredibly botched? Look here—see how the rods and cones are facing away from the light. *No Intelligent Designer would have made the eye that way!*"

"Two words of advice to my premed friend: Avoid *ophthalmology.*" Jackie picked up her cousin's pen and quickly shaded in a layer at the very back layer of the retina and then scribbled the letters "RPE" next to the layer. "Don't you know that if the vertebrate eye wasn't designed with the rod and cone tips embedded in the RPE layer of cells, then the vitreous humor would fill with all sorts of visual 'confetti'—sort of a dandruff in the eye? The eye would cloud up. Fortunately, the RPE gobbles up the debris, along with doing some other wonderful favors for the retina. Not botched design—it's great design!"

Kiera stared at the napkin. "Fine. So Miller and Dawkins need to tweak things a bit. What about the Therapsids?"

"The what-sids?"

"Therapsids. Mammal-like reptiles. The bones in their jaws get smaller and smaller and eventually they build into the inner ear. It's a perfect example of evolutionary transition."

"It's a big leap between a little jaw and an inner ear."

"At least admit that it's a possibility."

"It's a possibility...like it's a possibility that I could jump off a one-hundred-story building and not get killed. Oh, and no parachute."

"You saw that coming."

"And, as a biochemist wannabe, I know that mutations—the ones that are big enough to see—are almost always harmful."

"That's the beauty of it. The key mutations don't have to be large and visible. They just need to happen. And...they just so happen to have a lot of time to do so."

"Irrelevant! Even after one hundred and fifty years of bacteriological research, we still haven't seen the wholesale, macroevolution that your Darwinist heroes preach. At the risk of repeating myself, all we see *anywhere* is small-scale *microevolution* tweaking an existing structure."

"I'm not talking about one hundred and fifty years. I'm talking about billions of years."

"Bacteria only have a life span of hours or minutes, so one hundred and fifty years is more than enough time to witness macroevolution. But you know what? It doesn't happen!"

"You still haven't dealt with the clear cases of evolution, like the horse series or archaeopteryx or a hundred other perfect examples of Darwinian evolution."

"You've got to be kidding! Both of those examples are now suspect. And we haven't even talked about the number one problem: Evolutionists have yet to explain the biggest leap of all—how life came to be in the first place."

The Odds and the Faith

"I know Miller's experiments were somewhat flawed, but it's still an infant study."

"No matter. A hundred years from now the odds of the random organization of inorganic matter into the simplest replicating cell will still be, for all intents and purposes, zero!"

Kiera shook her head slowly. "Look. Evolution happened. The odds are irrelevant. Even the Patriots won the Super Bowl. *Sorry it doesn't jive with your religion.*"

"It's not about religion. It takes more faith to believe in Darwinism than it does to believe in an intelligent designer. Think of it this way. Your faith tells you that *strictly mindless forces*, with no guidance or direction, relying on random copying errors, took some pulsating blob of matter and turned it into a magnificent human being over time, with about sixty trillion cells, each carefully constructed in intricate array out of trillions of atoms, according to a blueprint whose complexity amazes our greatest scientists. My faith is that this wonderful complexity was put together by an intelligence..."

Kiera cut in, "But you're talking about *faith in the unseen*, and I'm talking about faith in the scientific method. Besides, what you're hinting at is a vast Darwinian conspiracy against science itself. That plain doesn't jive with mine or Gould's or Dawkins' honest investigation and conclusion that big-scale evolution is what has happened."

"No. I couldn't disagree more on where the evidence points. You hold to something that...what is it?...one hundred and fifty years later?...remains plagued by a lack of fossil evidence, the absence of any mechanism for generating the first cell, and Darwin's black box itself—the cell's complexity. And, yes, I do believe that Behe's evidence is the clincher—that certain machines cannot have developed by a random, successive series of mutations. They come as is, or the nonfunctional 'under construction' stages would not be able to survive."

"Fine! Then enlighten me as to your 'scientific alternative for future research,' O Great One. What shall we submit for research grants? Shall I tell the NIH that I need $500,000 for my project, 'Determining the theological implications of the flagellum'?"

"I suggest we start with a quick diagnosis: I won't use the word *conspiracy*. I'll just say that hundreds of generally diligent and honest scientists, because of their strong bias, shut out those views that run contrary to their naturalism. In a nutshell, *they're sure that the universe is all there is.* They're convinced of it. They know that any hint otherwise might eventually point to a designer, but they view any such idea as 'prescientific.' Lewontin at Harvard had the honesty to admit this—that Darwinists rule out divine action ahead of time, before the evidence is presented. They can't 'allow a divine foot in the door'!"

"Lewontin said that?"

"I'll get you the documentation. And the Darwinian elite certainly don't *prefer that there be a designer,* for several reasons I don't have time to go into. As a result, either knowingly or unknowingly, they shut out any mention of all such contrary views from their textbooks, and from just about all the public education in science, lest the ignorant public look into it and find their naturalistic explanations somewhat lacking."

"Again, you're hinting at religion."

"I guess I am. But who are we to try to silence the implications of good science? If our investigation keeps giving us hints of God, what are we to do with those hints?"

Kiera smiled impishly. "Lock 'em up and throw away the key."

Jackie grinned, sensing that Kiera was warming to new possibilities. "Did you know that Newton said that his whole thing about science was that it revealed God? Don't even get me started on Pasteur or Pascal or Einstein's belief in a godlike intelligence embedded in the universe."

"Well, maybe Behe's not a fringe wacko. Maybe they don't let dimwits teach at Lehigh. But I'm still committed to an evolutionary explanation of life. I still think the evidence leans my way."

Jackie responded gently. "I think that'll change. Darwin had no way to appreciate the complexity of God's creation, and now evolutionists are slowly being forced to come to grips with the implications of nonevolved design all around them—molecular and cellular structures. Physicists and astronomers are pouring out books and articles on the universe's own 'fine-tuning,' and it looks now like the fine-tuning even includes the makeup of the earth itself as an ideal scientific laboratory."

Kiera looked down, slowly twirling her coffee cup on the table. "I don't know how I feel about being *this* open-minded."

"Pardon the pulpit, but Jeremiah 33:3 says, 'Call to Me, and I will answer you, and show you great and mighty things, which you do not know.' Can your feelings, and do an honest experiment—just call out to God. He'll open up your mind to what He's done and what it means for you."

"Well, maybe I can blame Pascal and Einstein for intelligent design. But there's no excuse for bad coffee."

Jackie dug into her backpack for her change purse. "I think I can scrape up enough for Java Joe's."

"Your treat, Mrs. Behe."*

---

* The editors gratefully acknowledge the assistance of Jason Anderson in the writing of this chapter.

It is absolutely safe to say that if you meet somebody who claims not to believe in evolution, that person is ignorant, stupid, or insane (or wicked, but I'd rather not consider that).

—Richard Dawkins, prominent Oxford scientist and author

TWO

Chapter 2

Challenging Darwin's Myths

By Mark Hartwig, Ph.D.

Ever since Darwin first published his theory of evolution, his defenders' favorite tactic against critics has been to attack their character and intelligence. Darwin himself used it against some of the greatest scientists of his day, accusing them of superstition and religious bias. Now that Darwinism rules the scientific roost, such charges are widespread. California's science education guidelines, for example, instruct teachers to respond to dissenting students by saying, "I understand that you may have personal reservations about accepting this scientific evidence, but it is scientific knowledge about which there is no reasonable doubt among scientists in this field."

By today's rules, criticism of Darwinism is simply unscientific. Schools usually don't attempt to defend the theory against skeptics. A student who wishes to pursue such matters is told to "discuss the question further with his or her family and clergy." But is Darwinism so obviously true that no honest person could doubt it? Are alternatives so unscientific that no reasonable person could embrace them? The answer to both questions is a resounding no.

Searching for Support

The essence of Darwin's theory is that all living creatures descended from a single ancestor. All the plants, animals, and other organisms that exist today are products of random mutation and natural selection—or *survival of the fittest.*

According to Darwin, nature acts like a breeder, overseeing biological change. As useful new traits appear, they are preserved and passed on to the next generation. Harmful traits are eliminated. Although each individual change is relatively small, these changes eventually accumulate until organisms develop new limbs, organs, or other parts. Given enough time, organisms may change so radically that they bear almost no resemblance to their original ancestor.

Most importantly, this process happens without any purposeful input—no Creator, no intelligent designer. In Darwin's view, chance and nature are all you need. This all sounds very elegant and plausible. The only problem is, it has never been supported by any convincing data.

For example, consider the fossil evidence. If Darwinism were true, the fossil evidence should reveal lots of gradual change, with one species slowly grading into the next. In fact, it should be hard to tell where one species ends and another begins. But that's not what we find.

As Darwin himself pointed out in his book, *The Origin of Species:*

> The number of intermediate varieties, which have formerly existed on the earth, [must] be truly enormous. Why then is not every geological formation and every stratum full of such intermediate links? Geology assuredly does not reveal any such finely graded organic chain; and this, perhaps, is the most obvious and gravest objection which can be urged against my theory.

Darwin attributed this problem to the imperfection of the fossil evidence and the youthful state of paleontology. As the discipline matured, and as scientists found more fossils, the gaps would slowly start to fill, he thought.

Against the Evidence

Time has not been kind to Darwinism, however. Paleontologists have certainly found more fossils, but these fossils have only deepened the problem. What paleontologists discovered was not gradual change, but stability and sudden appearance. It seems that most fossil species appear all at once, fully formed, and change very little throughout their existence.

This poses quite a challenge for Darwinist paleontologists. One such paleontologist, Niles Eldredge, put it this way:

> Either you stick to conventional theory despite the rather poor fit of the fossils, or you focus on the [data] and say that [evolution through large leaps] looks like a reasonable model of the evolutionary process—in which case you must embrace a set of rather dubious biological propositions.

Large evolutionary jumps are anathema to good Darwinists because they look too much like miracles. Reptiles simply don't hatch birds.

The fossil evidence appears particularly troublesome with the "Cambrian Explosion," which paleontologists believe took place about 530 million years ago. In an instant of geological time, almost every animal phylum seemingly popped into existence from nowhere.

A *phylum* is a division that biologists use to classify the life on this planet. The organization starts with five kingdoms: animals, plants, fungi, protists, and bacteria. These kingdoms are then broken into *phyla*, which are divisions based upon more detailed characteristics. A phylum is, in turn, broken into class, then order, family, genus, species. Each level is more specific and less comprehensive than the preceding level until, at the species level, the division contains only one type of organism, for example "human being."

If the differences within a phylum are vast, the differences between phyla are really wild. As much as a chimpanzee may differ from a fish, it differs even more radically from a sea urchin. The two are built on entirely different architectural themes.

That's why the Cambrian Explosion remains so troubling for Darwinists. What paleontologists find isn't just the sudden appearance of a few new species. They encounter species so utterly distinct they have to be placed in completely different phyla. Even Oxford zoologist and prominent Darwinist Richard Dawkins has remarked, "It is as though they were just planted there, without any evolutionary history."

Worse yet, after the Cambrian Explosion, almost no new phyla appear in the fossil record—and many go extinct. By conventional dating, that's a 500-million-year dry spell. This is exactly the opposite of what Darwin would have predicted. According to Darwinism, new phyla are produced by the gradual divergence of species. As species split off from each other over time, they eventually become so dissimilar as to constitute a whole new body plan. Therefore, we should see new species slowly appearing over time, followed by the much slower appearance of new phyla—what Harvard paleontologist Stephen Jay Gould called a "cone of increasing diversity." Instead, the cone is upside down. Even by conventional timelines, the fossils look very non-Darwinian.

Darwinists, of course, express confidence that future discoveries will clear up the mysteries. But so far, the research has only deepened them. A recent reassessment of the fossils has added perhaps fifteen to twenty new phyla to the Cambrian zoo. Moreover, discoveries in 1992 and 1993 have shrunk the explosion's estimated duration from forty million years to about five million.

Science or Philosophy?

The fossil problem is only one of Darwinism's woes. Virtually every other area of research poses problems, too. But like the bunny in the Energizer battery commercials, Darwin's theory just keeps going. Why? Perhaps because Darwinism is more wishful thinking than fact.

Professor Phillip Johnson just finished a distinguished career as a professor of law at the University of California at Berkeley. While on sabbatical in England several years ago, he became fascinated with the serious problems in Darwin's theory. He was also struck by how Darwinists continually evaded these difficulties with tricky rhetoric and pulpit pounding. As he dug deeper into the scientific literature, Johnson eventually became convinced that Darwinism wasn't so much a scientific theory as a grand philosophy—a philosophy that attempts to explain the world in strictly naturalistic terms. "The whole point of Darwinism is to explain the world in a way that excludes any role for a Creator," says Johnson. "What is being sold in the name of science is a completely naturalistic understanding of reality."

According to Johnson, the reason Darwinism won't die is that its basic premise is simply taken for granted: namely, that chance and the laws of nature can account for everything we see around us, even living things. Given that assumption, Darwinism has to be true because nothing else is permitted inside the arena of possible explanations. Creation has been ruled out from the start, and the other naturalistic theories are worse than Darwin's. So any argument against Darwinism is usually ignored.

Ruling Out Design

Today a new breed of young scholars is challenging those Darwinist assumptions. Their ranks include biochemists, microbiologists, astronomers, and philosophers. They all argue that design is not only scientific, but is also the most reasonable explanation for the origin of living things. And they're gaining a hearing.

One such scholar is Stephen Meyer, a graduate of Cambridge University in the philosophy of science and now a professor at Whitworth College in Spokane, Washington. Like Johnson, Meyer believes that the prohibition of design has essentially stacked the deck in favor of Darwinism. "There's been a kind of intellectual rigidity imposed on the origins discussion," Meyer says. "It's only possible to talk about origins in a naturalistic vein, because people believe that the rules of science prohibit talking about Intelligent Design."

This prohibition rests on what philosophers call *demarcation standards.* These criteria allegedly set science apart from other disciplines such as theology, history, or literary criticism. For example, someone might say that a scientific theory must explain everything in terms of observable objects and events, that it must make predictions, or that it must be capable of being proven wrong. "Although scientists and philosophers have proposed many demarcation standards," says Meyer, "none of them do what evolutionists want them to—which is to exclude Intelligent Design as a scientific theory. When applied evenhandedly, demarcation standards either confirm that design is scientific, or they exclude evolution, too."

Again, for example, Darwinists like to argue that design is unscientific because it appeals to unobservable objects or events, such as a Creator. But Darwinism also appeals to unobservables. "In evolutionary science you have all kinds of unobservables," Meyer says. "The transitional life forms that occupy the branching points on Darwin's tree of life have never been observed in the rock record. They've been postulated only because they help Darwinists explain the variety of life forms we observe today."

"When scientists try to reconstruct past events, appealing to unobservables is entirely legitimate," Meyer says. What's illegitimate is to say that design theorists can't do the same thing.

Design As Science

William Dembski, another Evangelical scholar, is a research professor at Baylor University in Waco, Texas. He holds a Ph.D. in mathematics from the University of Chicago and another in philosophy from the Chicago campus of the University of Illinois. He recently wrote three books: *The Design Inference* (published by Cambridge University Press), *Intelligent Design*, and *No Free Lunch*. In these books he has set forth the basic principles of a whole new system of thought and probability measures by which scientists can now "detect design."

Dembski argues that Intelligent Design, far from being a strange and exotic notion, is something that science recognizes every day. The existence of entire industries depends on being able to distinguish accident from design—including fraud investigation, criminal justice, cryptography, patent and copyright protection, and many others. No one calls these industries "unscientific" simply because they look for evidence of design. Indeed, some scientific disciplines, such as anthropology and archaeology, could not exist without the notion of intelligent design. "How could we ever distinguish a random piece of stone from an arrowhead except by appealing to the purposes of primitive artisans?" asks Dembski.

According to him, we recognize design in events or objects that are too improbable to happen by chance. Stones don't turn into arrowheads by natural erosion. Writing doesn't appear in sand by the action of waves. An unaltered coin doesn't come up heads a hundred times in a row. Such results point to some intelligent cause.

There's more to design than low probabilities, however. If someone tosses a coin a hundred times, duplicating any string of results will be extremely improbable. But if someone claims that the coin came up heads a hundred times in a row, we would suspect that something more than chance was involved. "Our coin-flipping friend who claims to have flipped a hundred heads in a row is in the same boat as a lottery manager whose relatives all win the jackpot or an election commissioner whose own political party repeatedly gets the first ballot line," Dembski says. "In each instance public opinion rightly draws a design inference and regards them guilty of fraud."

If detectives can use this kind of thinking to spot election and lottery fraud, and if archaeologists can use it to spot arrowheads, then why can't biologists use it to look for design in the living world?

Irreducible Complexity

Even without precise definitions, it's not hard for most of us to recognize design in the living world. The exquisite complexity of living organisms virtually proclaims the existence of a Creator. Many Darwinists admit this—except they say it's only an illusion, produced by strictly natural forces.

For Michael Behe, a biochemist at Lehigh University in Bethlehem, Pennsylvania, the complexity is too extreme for Darwinism to be plausible. He argues that many systems in living organisms are irreducibly complex. They consist of several parts, all of which must be present for the system to work. "It's like a mousetrap," says Behe. "A standard household mousetrap has five parts, all of which must be present for the trap to work. If you take away any of those five parts, you don't have a functioning mousetrap. You can add the parts one by one, but until you get to the full five parts, you have no function. It's an all or nothing kind of thing."

This irreducible complexity exists even at the level of a single cell. "It was originally thought in Darwin's day that cells were very, very simple things—like little blobs of gel. But as science has progressed, it's shown that cells are extraordinarily complex, more complex than anybody thought."

One example is the system that transports proteins within the cell from where they're made to where they're used. *Enzymes* are a class of proteins that help the cell digest other kinds of proteins. They are created in a compartment called the *endoplasmic reticulum*. But they do all their work in another compartment, the *lysosome*. To get from the one compartment to the other, enzymes are stuffed into a *vesicle*, a kind of bus. The "bus" then travels to the destination compartment and eventually merges with it, spilling its contents into the compartment. Achieving this task requires several very specific proteins. A cell needs certain proteins (along with certain fats) to form the little capsule that contains the enzyme. It needs others to help the capsule grab the right protein. Finally, it needs proteins that help the "bus" attach itself to the destination compartment and merge with it.

"Now if you think about irreducible complexity," Behe says, "virtually all of these proteins have to be there from the beginning or you simply don't get any function." That makes it tough for Darwinists to argue that design is simply an illusion that has been produced by mutation and natural selection.

"Darwin said one thing pretty strongly in *The Origin of Species*," Behe notes. "He said that if it could be shown that any system or organ could not be produced by many small steps, continuously improving the system at each step, then his system would absolutely fall apart. Now the thing about irreducibly complex systems is that they cannot be produced by numerous small steps, because one does not acquire the function until close to the end, or at the end. Therefore, with irreducibly complex systems, they cannot be produced by Darwinian evolution."*

---

* Editor's Note: The contribution of Dr. Behe is so important, we have included two chapters on his work: "Meeting Darwin's Wager" by Thomas E. Woodward and "Modern Intelligent Design Hypothesis" by Michael J. Behe.

Gaining Ground

Most scientists are still far from throwing in the towel on Darwinism and accepting intelligent design. Nevertheless, design advocates are finding it easier to gain a hearing. One of the most important of these hearings took place in March 1992 at Southern Methodist University in Dallas. In this landmark symposium, Phillip Johnson, Stephen Meyer, William Dembski, Michael Behe, and other Christian scholars squared off against several prominent Darwinists. The topic of debate, "Darwinism: Science or Philosophy?," was suggested by Johnson's own criticisms of evolutionary thought in his book *Darwinism on Trial.*

The remarkable thing about the symposium was the collegial spirit that prevailed. Creationists and evolutionists met as equals to discuss serious intellectual questions. Not surprisingly, few issues were resolved. But in today's Darwinist climate where dissent is frequently written off as religious bias, just getting the issues on the table was quite an accomplishment. What's more, several months later, one prominent Darwinist who participated in the Dallas symposium publicly conceded that one of the points Johnson made was correct—namely, that Darwinism is based as much on philosophical assumptions as on scientific evidence. This admission took place at a national meeting of the country's largest science society, the American Association for the Advancement of Science. It scandalized the Darwinist community, which likes to portray evolution as an indisputable fact. It was all the more scandalous because the speaker had specifically been invited to the meeting to denounce Johnson.

Design advocates are still far from winning, but they believe things are getting better. As Johnson points out, Intelligent Design arguments are getting more sophisticated, while most Darwinists are still responding with clichés. Now it's the new creationists who come across as asking the hard questions and proposing bold solutions to persistent scientific mysteries.

But ultimately, Johnson says, it's not the debates or the arguments that will win the day. "It's reality that's doing it. It's just the way the world is. And sooner or later, scientists will have to acknowledge that fact."*

*"Challenging Darwin's Myths" first appeared in *Moody Magazine*. It is reprinted by permission of the author, Dr. Mark Hartwig.

Mark Hartwig

In the late 1970s, a bright group of university students in California began to ask probing questions about the plausibility of Darwinian evolution. As they uncovered the rapidly crumbling foundations of Darwinism, they decided to organize research into this area and launched "Students for Origins Research," which later became the Access Research Network (ARN).

One of the brightest and most capable of these university graduates who played a strategic role in launching ARN was **Mark Hartwig.** He received a Ph.D. in educational psychology from the University of California at Santa Barbara. Dr. Hartwig is one of the early organizers of the Intelligent Design movement, and his articles have appeared in the *Wall Street Journal*, the *Denver Post*, the *Los Angeles Times*, and several other national newspapers and magazines. He was the editor of the journal *Origins Research* and contributed to the revised edition of the widely used textbook *Of Pandas and People.* He now does research and writing on public issues for Focus on the Family.

What is the earth most like?...It is most like a single cell.

—Lewis Thomas in *The Lives of a Cell* (1974)

The Magnificently Complex Cell

BY JAMES P. GILLS, M.D.

The living cell is, in itself, an entire universe of molecular suns and planets in practically perpetual motion. It is the basic brick of life's awesome, myriad constructions from giraffes to jellyfish, the essential beating drum of all the rhythms of nature's exquisite dance, and considering that many of God's creatures are unicellular, it is the virtual definition of life itself. The cell is far more complex than any large, modern city that we are familiar with or any celestial galaxy that we gaze upon. Human science has just begun to plumb the profundity of this miraculous cell. To encounter its inherent complexity is to perceive a design that is as surprising and inexplicable as the fortuitous discovery of a priceless wristwatch on barren sand.

The Living Cell

Reasonable thinkers would immediately presume such a watch to be the work of a designer, an *intelligent Designer.* However, for more than a century, the presiding tenet in the educated circles that consider our collective origins has been the explanatory principle of Darwinism, simply stated: Happenstance imperfections in the replication or copying of genes, and the demands of specific environments, *gradually* produced the observed beauty, synergy, and ingenuity of cell design and function—and of life. Yet details that we are only recently becoming aware of make the notion of the cell's evolution out of nothingness, by chance alone, more absurd even than the proposition that the fine Swiss timepiece in the dune arose spontaneously on that spot.

A new view

Prior to my own introduction to cell biology in medical school, Sir John Eccles had recently been awarded the Nobel prize in medicine for his work on the cell and its electrophysiological transmembrane potential. As a student at Duke University, I devised a "nanopipette" to measure just such electric potentials across a single cell membrane. Eccles had exposed the cell as an active and intricate entity possessing dynamic relationships between a number of cell systems. This was new and provocative. By this time as well it was becoming clear to embryologists that, appearances notwithstanding, growing fetuses of any species were in fact not passing through the stages of a long, shared evolutionary history as the classic "ontogeny recapitulating phylogeny" postulate had asserted.

My personal experience at this time was a bright awakening to a spectacular complexity that I had not confronted previously, compounded by a growing sense of the inadequacies of Darwin's theory of evolution of species. Based on detailed observation of living things, I had come to feel the intellectual tension between polar opposites tugging me more in the direction of Creation and away from popular theories of evolution.

Eccles's work had begun to cause the foundations of commonly cherished theories of the cell to quake. The same cell had been little more than a mere glob in Darwin's mind as he propounded the theories that would hold biology in thrall to this day. Behe has stated this: "At the tiniest levels of biology—the chemical life of the cell—we have discovered a complex world that radically changes the grounds on which Darwinian debates must be contested."[1] Darwinian evolution, Charles Darwin's own assumptions, neglected to...could not at the time...account for the microscopic universe of the cell and thus stood on shifting sands rather than solid ground. Professor Michael Behe's seminal book has labeled the magnificently complex cell as "Darwin's black box"—an object impenetrable by light or wisdom to Darwin and his contemporaries.[2] While the light is now, in our time, reaching the inner sanctum of the cell, for many individuals the wisdom has yet to do so.

Current, rapid advances in molecular biology and other disciplines have uncovered a complexity in nature that conventional methodology falls short of explaining. Still, though the details of biology, chemistry, mathematics, physics, geology, and astronomy evidently adduce the role of a Creator, the egoism of many intellectuals, of many people, leaves God on the periphery of experience as either mere opiate or crutch for the weak, rather than as universal architect.

The abiding paradox that I have observed in many professionals, including some Nobel laureates, is that the acknowledgment of the intrinsic complexity of the cell has failed to impel them to credit a Creator. These are brilliant thinkers blinded by *spiritual cataracts,* left groping for answers by the *spiritual myopia* that human pride afflicts. We boast of human intellectual development as a universal zenith on the one hand, while on the other we neglect to note our failure to understand fully the working of even one cell in the body. And as the complexity incrementally increases, our inability to grasp the whole deepens, with the integration of cells into organs, and finally organs systemically and synergistically integrated into a complete human body. And what a body it is!

Physical testament

Richard Swenson's fine book *More Than Meets the Eye* is recommended reading as a potent alternative to Darwinism. In rich detail it expertly provides an in-depth examination of the attributes and assets, some of which follow, that make us what we are, and how we came to be so:

- One human body contains more atoms than the universe does stars.[3]
- We each have nearly 60,000 miles of blood vessels.[4]
- During their life, the average person will breathe 630 million times, inhaling approximately 22 pounds of air per day.[5]
- We blink our eyes over 400 million times before we close them permanently.[6]
- The human eye, with its 120 million rod cells and 7 million cone cells, can distinguish millions of differing color shades and can process information faster than any existing or imagined supercomputer.
- The 10^{13} cells that comprise the average human body initially derive from a single, solitary cell.[7]
- One cell will not only divide, but, of great mystery, differentiate into over 200 different types, each with a unique and specific role.

Could man or mere time and accident be responsible for this marvel? Is it not pride that occludes our perception of structure as God's structure, and complexity, integration, synchronization, and rhythm as His design?

Your own eye offers a particularly useful illustration of the absurdity of Darwin's theory when it is applied to the incredible bodies that we inhabit. The human eye presents a conundrum for those theorists trying to explain where the distant, more primitive precursor to the eye was. As a unit, the human eye represents what Behe has referred to as an "irreducibly complex" system. Such a system is "composed of several well-matched interacting parts that contribute to the basic function, wherein removal of any one of the parts causes the system to effectively cease functioning."[8]

Darwin would be astonished by the wealth of details that we now possess, and he would be forced to reconsider his basic thesis in the face of them since the small, successive, progressive modifications that gradual evolution predicts could not lead to such an irreducibly complex system such as the human eye. This is explained by the fact that any precursor system, by definition, would be missing some part or other, and that system necessarily could not function without that part. An irreducibly complex system would be forced to announce its presence—if it is to function—in one step. Here, the Swiss watch in the sand? Here, the hand of a Designer.

My epiphany in medical school led me to understand that although we could look at cells and study them, we could neither create them nor fully comprehend them. Nobel prize-winning biologist Alfred Gilman has admitted, "I could draw you a map of all the components of a cell and put in all the proper arrows connecting them...but...I or anyone else would look at the map and have absolutely no ability to predict anything."[9] We must remain humble before, and as, God's creation.

As a Christian, Eccles believed that it is important to question the claim of science to sole possession of final truth. It should be added that, in a similar light, we need to revise the orthodox approach to biology given by a mechanistic theory that reduces life to mere principles of physics and chemistry.[10] The stunning, bewildering complexity of the machinery of life, the testimony of an intelligent design, from our cells to our smiles, ought to leave us awed and humble in a search for truth that reaches deeper than the petty concerns and preoccupations that too often engage us. The details of our constitution, the example of the cell, refute Darwin's dogma and point to God's truth.

Wonders of the Cell

Six billion people share this earth, and every one of us, just as those innumerable souls that have gone before us, began the adventure with a single, microscopic egg opening the doors to a single sperm cell. Those two cells, each possessing only half of the genetic complement that makes an individual unique and special, meld into a singular, wholly original cell that starts to perform magic, starting the symphony. The fertilized egg will begin to divide into a colossal 10 to 100 trillion cells—each cell comprised of approximately a trillion atoms. All of these cells, by the time you are born, will have differentiated into over two hundred varieties, with specialized, prescribed jobs to faithfully attend to—liver cells, kidney cells, brain cells, everything that makes us what we are. Each cell expresses an extraordinary and incomparable complexity, beauty really, of form and function. This is a complexity of "multiple interwoven requirements" that a dictum of gradualistic evolution cannot account for.[11]

Some cell anatomy

If we approach a cell from the outside we initially encounter the cell membrane, the "skin" of the cell, which is more diaphanous and delicate than the tiny filament of a spider's web.[12] This membrane is more than practical encasement; it is an active, vital, and busy border. The cell generates an electrical field; the potential, which I measured as a student, can be, at times, larger than the electrical field found near power lines.[13] The ostensible task conducted at this border is that of inviting in and absorbing molecules of nourishment like glucose, oxygen, and proteins from the interstitial fluid that surrounds and bathes the cell. Conversely, the same membrane jettisons undesirable waste products from the cell's busy interior into that exterior interstitial fluid. This dedicated activity facilitates the task of the "hearts" of the cell, the mitochondria.

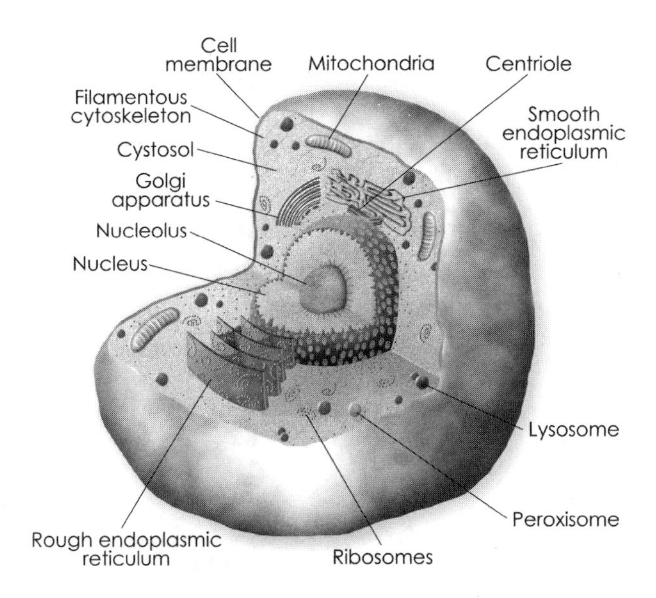

A Typical Human Cell

Tiny organelles floating in the cell's internal lake of cytoplasm, mitochondria are infinitesimal power plants, each with hundreds of little factories called *biofires* embedded in its sides. Each biofire, two hundred thousand times smaller than a pinhead, earnestly produces a fuel molecule called *adenosine triphosphate* (ATP), which the cell uses for energy. Active people can produce their body weight in ATP every day![14] As with the consumption of any fuel, waste, carbon dioxide for example, is produced and must be dispatched outside the cell to be expelled eventually from the body via kidneys, liver, and lungs. None of this toil is heedless or random—it is all conducted under the direction of the cell's internal clock, which switches on and off in unvarying cycles of between two and twenty-six hours, depending on the cell-type in question.[15] The mitochondria share their cytoplasmic home with the "brain" of the cell, the nucleus.

The nucleus contains the twenty-three pairs of chromosomes, one pair from mom and one from dad, by which each human is physically defined. These chromosomes are parcels of coiled deoxyribonucleic acid (DNA), "perhaps the most spectacular of all human miracles."[16] Organized as stretches, or sentences, of code called *genes*, this DNA—*your* DNA—will determine not only your humanness, but also your eye color, the amount of curl in your hair, the sound of your laughter, and, likely, much of your personality. It is the genome, the gene code in its entirety, with all its inherent secrets of the individual and the species, that was the object of the highly publicized mapping endeavor called the Genome Project.

DNA's blueprint

In 1953 James Watson and Francis Crick proposed the model of the double helix DNA molecule and were rewarded with a Nobel prize for their work. This now famous molecule consists of two strands of sugar and phosphate, which wind about each other along their length. Four nitrogenous bases—adenine, thymine, cytosine, guanine (A, T, C, G)—are arranged along the inside of the twisting sugar-phosphate backbones forming

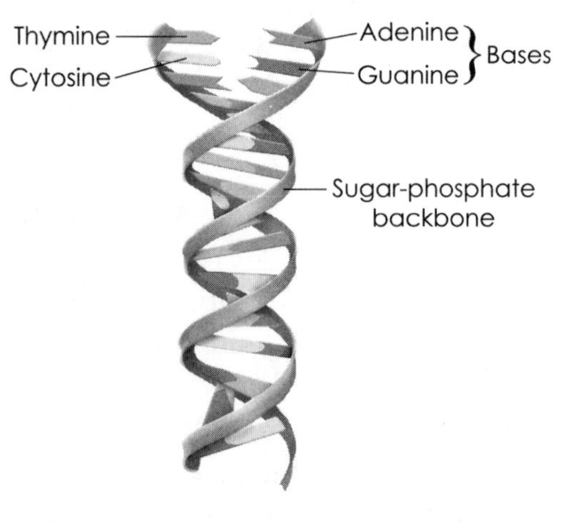

The DNA Molecule

pairs that correspond, if you will, to the rungs of a spiral ladder. Held together by hydrogen bonds, A pairs with T, and C with G in a particular order. It is this order of bases that encodes all the genetic instructions of an organism.[17] Such an apparently simple structure betrays an immense complexity.

- The DNA from a single cell, stretched out, would extend over five feet in length.[18]
- One strand is only 50 trillionths of an inch wide.[19]
- Removal of DNA from all of one individual's cells and stretched out from end to end would extend in a strand from a minimum 10 billion miles to a possible maximum of 170 billion miles.[20]
- A strand of DNA stretched from you to the sun would weigh a slight 1/50 of an ounce.[21]

This is an astonishingly delicate and yet profound filigree that contains vast stores of information. The information, the genome, makes proteins, and as Swenson has noted, the proteins "do everything."[22] Reading the genome sequence at a rate of ten bases per second, nonstop, would require nine and a half years to complete. The human genome project was a very heady task from the outset, yet is it not audacity to presume that such a wonder as our genetic blueprint could be fully apprehended? Could it really have arisen by accident?

Recent evidence from the genome project provides even clearer indications of design within the mystery of DNA. The human genome is now thought to be smaller than was originally assumed, and current suggestions are that it possesses only a few thousand more genes than a mustard weed plant.[23] However, the human genome is "inexpressibly elegant" and efficient:

> In many organisms, each gene codes strictly for a single protein, but in humans...that single one-to-one ratio

> doesn't hold. Human genes interact with one another in a variety of intricate arrangements; as a result, a single human gene is often responsible for producing a handful of proteins or more.[24]

We find *inexpressible* elegance and efficiency within us. Words fail us mightily in our attempt to explain and describe the miracle. Are we not myopic when we view ourselves within creation? The framework of our life is given to us by the cell and its DNA. In this single, truly perplexing cell is the irreducible complexity that attends to a beautiful, intelligent design and Designer. Often, when confronted with the spiritual cataracts of those people who refuse to see the artistry of God in our physical constitution, I inquire whether they have taken the time to consider their own pancreas.

Man may be capable of mapping the genes in a pancreatic cell, but man cannot create or duplicate the integration and synergy exhibited by the cellular lobules, organ ducts, and secreted enzymes all in communication with the liver, digestive system, and the body as a whole. The islets of Langerhans in the pancreas alone have the daunting job of creating the insulin that regulates blood sugar levels, and then secreting it in appropriate amounts at the correct time. Such integration of cells, glands, fluids, and organs is staggering, and still we refuse to look to a higher mind than our own. The symphony that is life, each note played faithfully by the individual cells in their section of the orchestra, must be conducted by an awesome Creator. Without Christ we are doomed by our superficiality, our myopia, our arrogance, and our pride.

Overview of the tissue repair process

The repair and regeneration of damaged tissue is a specific series of ordered reactions that is a remarkably potent manifestation of irreducible complexity and a Creator's intelligent design. The concerted, coordinated action of many specialized cells and the integration of their systems are functional requirements. Cell migration, proliferation, and attachment are key regulatory elements of a process that is directed by growth factors, themselves dependent upon appropriate receptors for proteins and appropriate signaling pathways to which the receptor is coupled. Significantly, growth factors belonging to one cell may have no effect without *multiple* factors being present. Again, this process of repair demonstrates a high level of specificity that is irreducibly complex.

Though healing begins at the edge of a wound, new tissue must form at the site of loss, not in surrounding undamaged tissue. The metaphor of immediate response toward, and reconstruction of, a bombed building is an apt illustration of what goes on in tissue repair. Within the first five minutes after trauma, platelets are hustled by the bloodstream to the scene of damage where coagulation in the presence of fibrin begins to stanch the blood flow. You may imagine this as the rapid response of fire engines and crews arriving to extinguish ravenous flames minutes after the initial incendiary event at the building.

Within one to six hours, *polymorphonuclear leucocytes*—white blood cells—are mobilized to the site to fight bacteria and remove debris much as ambulances would soon arrive at our allegorical building to start digging in the rubble and removing the injured. Then macrophages, the third tissue response, present themselves within twenty-four to forty-eight hours to remove clotted blood and damaged tissue in the way front-end loaders would arrive at the building site to clear the broken walls and damaged foundation.

Next, fibroblasts bring their ability to manufacture collagen, which is a major constituent element in all parts of the body, to the area; these are our metaphorical architectural planners and construction engineers. That stage invites the next phase of reconstruction with the introduction of the vascular endothelial cells, upon which fall the burden of *angiogenesis*—the formation of new blood vessels. With the task of delivering nutrients to and removing waste from the new tissue, these vessels are equivalent to the water and sewage lines for the new building. And then the body starts the bricklaying by stimulating mitosis, the division of cells into daughter cells.

Intricate, beautiful, and replete with many interconnected dependencies that display the irreducible complexity of an intelligent design, the tissue repair process shares the faculty of demonstrating God's handiwork with a variety of other specialized cells.

Specialized cells

During its one-hundred-twenty-day life span, an *erythrocyte*, commonly known as a red blood cell, speeds through a canal system of veins and arteries, which is approximately equal in length to two and a half times around the world at the equator.[25] This dimpled cell piggybacks the complex, iron-rich hemoglobin molecule to facilitate transportation and deposition of indispensable oxygen to all the body's hungry cells. Ingeniously, the red blood cell will normally release only about 25 percent of the oxygen that it is carrying to allow it a margin of potential in the event of abrupt, heavy oxygen demand, should you decide to run up the stairs.[26]

Consider also the wonders of the *myocyte* muscle cell of the heart. Under edict of nerve impulse it will contract just sufficiently to expel all the blood in a heart's chamber. It responds to excitement and adrenaline by beating faster and, as anyone on vacation knows, responds to calm by beating at a less rapid rate.

The *Kupffer cell* is unique to the liver and is designed as a phagocyte, a consumer, to ingest and break down matter that is toxic in the body. To understand Kupffer cell function, it must be considered in its context with other liver cells. The hepatocytes in a liver lobule form a series of plates only one cell thick, arranged like spokes of a wheel. All exposed hepatocyte surfaces are covered with microvillae—"hairs." Sinusoids, or spaces, between adjacent plates empty into the central blood vein. The walls of the sinusoids contain large openings that allow substances, the toxins, to pass out of blood circulation and into the spaces surrounding the hairy hepatocytes. The hungry Kupffer cells line these sinusoids and engulf to destroy pathogens such as bacteria, damaged blood cells, and heavy metals or other toxins absorbed by the digestive tract. As a collective, this represents an efficient, effective, and marvelous blood filtering system.

A body sentinel with superior programming is the hunter-killer *lymphocyte*. Like wasps, these cells are built for exploration and roam the body, sensing and monitoring, searching for molecular configurations of targets. The truly amazing aspect of these insatiable cells is that they will fit with polymers that do not yet exist until some chemist in some lab has synthesized them. In other words, these cells "do more than predict reality; they are evidently programmed with wild guesses as well."[27] This is mysterious, and evidently indicative of deep design.

Dr. James Rowsey provides us with another excellent example of specialized cellular structure and function with his description of the human eye's *endothelial cell of the cornea*. It was the first cell in living patients observed to respond to various disease states in the body. An *endothelium* is a layer of cells lining the innermost surfaces of many organs, glands, and blood vessels. As Dr. Rowsey explains, the eye's cornea has approximately thirty-five hundred endothelial cells per square millimeter, which provide both a barrier to water flowing into the cornea and a water pump from the cornea into the anterior chamber of the eye. Surprisingly, these endothelial cells are able to recognize minute trauma on the surface of the cornea, such as a contact lens abrasion or even contact lens anoxia—oxygen deficiency—by widening the junctional complexes that face the anterior chamber and allowing increased fluid flow into the cornea, bringing with it additional nutrients to preserve the integrity of the corneal tissue. The endothelial cells, endowed with nature's most geometrically and thermodynamically stable hexagon shape, breathe oxygen on the outside of the cornea, and with every blink this new oxygen is supplied to the aqueous component of the tear film.

These cells have an uncanny ability to recognize and respond to distant trauma and disease, even at a distance of three or four millimeters away. This is comparable to a parent who is living in Seattle hearing her child cry in Washington, D.C. When there is damage, the cells recognize the problem, for example a corneal ulcer, and begin to regenerate a new Descemet's membrane on the back surface of the cornea as a *preemptive* mechanism aimed at preventing the cornea from melting away and perforating. Once the cornea perforates, the eye is lost, so the endothelial cells recognize trauma and degradation at a considerable distance and get one step ahead of it. This is accomplished through the communications of inflammatory mediators notifying both keratocytes and the endothelial cells of the ongoing inflammatory cascade. You can think of this as a family unit in which all cells take notice when one of their collective number is injured, and then initiate a reparative series of actions. Would you be inclined to impute accident or design in such a remarkable system? We must be willing to search further into the mystery and to admit the existence of design when we are presented with it.

The *retinal photoreceptors* of the eye allow us additional glimpses into the magnificence of the single cell. These remarkable cells demonstrate both grandeur and design, as their description by another colleague, Dr. Dan Montzka, indicates. These remarkable cells comprise the outermost layer of the retina, which is a transparent, paper-thin structure that lines the back wall of the eye. The function of each photoreceptor cell is to capture light, photon by photon, and convert it into electrical signals that can be transmitted to the brain. Obviously, this is a complex series of reactions, and it can occur more than thirty times a second in each cell. With function over a range of ten million times difference in light intensity, these cells, by the demands of their extremely difficult job, have the highest energy requirements of any cell type in vertebrates like us! A single defect in any of the many cellular proteins can lead to a blinding disease such as retinitis pigmentosa.

There are two types of photoreceptors in the retina—rods and cones. Rods have a rodlike shape and are required primarily for night and peripheral vision. There are 120 million of these distributed throughout the retina of one eye. Cones have a cone-shaped outer segment, and we use these for color vision and the perception of fine detail. There are approximately 7 million cones per eye concentrated largely in the central retina. Each cone contains one of three color pigments that respond to differing wavelengths of light. This provides us the ability to discern colors. Again, a single DNA mutation can cause a color-blinding defect.

The astounding details of the multifarious forms and multidimensional functions that we observe in all these cells proclaim loudly that Darwinian theory is not based on testable, scientific fact, but rather upon the enforced prejudices of naturalistic assumptions. Challenged by the testimony of the magnificent cell, we must have the will and wisdom to sound the deeper levels of existence and, by stripping away petty, human pride, acknowledge the signs of God's intelligent design and construction in us, right down to our component parts. The details now available to us show Darwinism to be a fallacy, and also to be deadening to both scientific and spiritual insight as it deflects minds and hearts from the truth. The proof is in the details.

Darwinism Refuted

The essence of the Darwinian argument on origins is that *random mutations*, random changes in DNA acted upon by environment over enormous expanses of time, lead to minute advancements in an organism that eventually add up to wholesale changes in complete systems and morphologies. The essence of the design argument is that a designed system would cease functioning if one of several component parts were missing, and so, gradual, minute alterations could not lead to a fully functional, complex system. Try as they might, no evolutionary theorist has yet been able to explain adequately how a watch, an eye, or healing skin, as systems of irreducible complexity, "might be produced without a designer."[28]

Despite such glaring explanatory shortcomings, Darwin's

adherents have remained dedicated because it is the only way they can describe the emergence of life without reference to a nonmaterial designer—to God. In truth, not only is Darwinian theory inadequate in explaining what we see, the notion that you, the reader, got here through a long process of random mutations requires a longer leap than "the leap of faith required to consider the possibility of life emerging from the work of an intelligent Designer."[29]

At the very beginning

Scientists for years have posited that life originated...was thrust up...in an ancient atmosphere and sea rich in hydrogen, methane, ammonia, and water vapor. With atmospheric lightning stirring the pot, amino acids, the building blocks of genes and proteins, coalesced in this prebiotic (before life) or primordial "soup," eventually aggregated, and began assembling primitive life. Orthodox biology has always fallen back on the argument of time, tremendously long stretches of it, to explain this process. We are now aware, however, that there has not been anything *close* to sufficient time.

Proponents of evolution were enraptured when, in December 1952, Stanley Miller's laboratory began to artificially produce amino acids in an experiment that simulated the prebiotic sea and its surrounding conditions on earth. But it is apparent that random conditions were not maintained. Behe points out that while any undergraduate with a manual can produce a piece of DNA in a modern lab setting, "there were no chemists 4 billion years ago, neither were there any chemical supply houses," leading to the conclusion that at the genesis of life, some intelligence was involved in the chemical reactions.[30] This conclusion is genuinely unavoidable because "a law of nature could not alone explain how life began, because no conceivable law would impel a legion of atoms to follow precisely a prescribed sequence of assemblage."[31] Mathematics supports this.

Probability refutes Darwin

Many have calculated the chances of accidental origins of life in a soup and its eventual diversification into you, your dog, and your potted violet. And the numbers are truly astonishing. They point to an intelligent design and a Designer, away from the naturalistic explanations of Darwinism.

The solution of the popular Rubik's Cube offers a colorful illustration of the impossibility of our *accidental* genesis by approximating the odds of the random evolution of a single protein of the human body. If a human subject were handed a cube and blindfolded, to ensure that all moves were random, this subject would need 1,350 billion years, at a rate of one move per second, to solve the cube.[32] That span is 300 times the accepted age of the earth that dedicated evolutionists allow in their 15-billion-year-old universe. The odds against each move producing a color match on the cube are $5^{1(19\text{zeros})}$ to 1.[33] Such odds are unfathomable given that we use more than 20,000 different proteins in our cells.

Our enzymes serve to refute Darwin as well. There are some 2,000 of these catalysts in our body, and the chance of finding all 2,000 by accident "is about the same as the chance of throwing an uninterrupted sequence of 50,000 sixes with an unbiased dice!"[34] There are an estimated 10^{80} atoms in the universe, and our odds for random emergence of all enzymes are 1 in $10^{40,000}$. This means that life could not have appeared by "earthbound random forces even if the whole universe consisted of primeval soup."[35]

Odds such as this are tantamount to impossibility, yet, unbelievably, people persist in their refusal to look for the Creator. The odds of life having been assembled by a designer are indeed greater. This enzyme statistic is a meaningful one in the context of this argument, because all 2,000 of them must form in an *exact way* for an organism to operate. It is this high level of interdependence of components, each alone hugely improbable as a random event, that gives us our evidence of intelligent design and banishes Darwin's theories as a historical artifact.

The odds are against even the spontaneous formation of rudimentary or simple forms. The probability of an *E. coli* bacterium, an uncomplicated cell in relative terms, arising in the prebiotic soup over a period of 5 billion years has been estimated at 1 in $10^{10(110)}$.[36] So it has wryly been stated that "the chances that life just occurred are about as unlikely as a typhoon blowing through a junkyard and constructing a Boeing 747."[37] It does not take a mathematician to conclude that spontaneous and random evolution is not only improbable but impossible.

Even Carl Sagan, renown, perhaps, for his atheism as well as for his science, estimated the difficulty of the chance evolution of a human at $10^{-2,000,000,000}$.[38] Can we really continue to turn a blind eye to these numbers? They speak of, perhaps bellow, something more. Within this heart of explanatory deficiencies of Darwinian theories of chance is the evidence of grand design, of irreducible complexity, in nature.

In that regard, the parable of two watchmakers is a particularly effective demonstration that turns the balance of interpretation in favor of irreducible complexity in design. Two watchmakers, Hora and Tempus, are each engaged in making watches of one thousand pieces. Hora assembles his watch one piece at a time. If there is a disturbance on his table, he must start from scratch once again. Alternatively, Tempus constructs sub-assemblies of ten parts each, then ten of these into sub-assemblies of one hundred, and finally ten of those to make the watch. If there is a disturbance at Tempus's workbench, he must repeat, *at most*, nine assembling operations. Hora countenances at least the *possibility* of redoing 999 operations.

At a hypothetical rate of one disturbance per one hundred operations, Hora will take four thousand times longer to assemble a one-thousand-part watch. "If, for mechanical bits we substitute amino acids, protein molecules, organelles, the time-scale becomes astronomical."[39] The support for irreducible complexity and intelligent design in this universe is not only overwhelming, but also self-evident.

The end of Darwinism

Darwin's theories have become obsolete under the weight of scientific data about the organization of structures at a molecular level, then systems, and finally life. Some evolutionists have more recently altered Darwin's primary thesis of step-by-step modification by positing long, quiescent periods of time followed by sudden leaps of development. Regardless of the differences, all versions depend on alterations in the genetic code and therefore have a fundamental, inescapable problem: From where did the genetic material derive?[40]

In medical terms, Darwinism is the equivalent to inheriting an error, a mutation of the DNA. In my capacity as a physician I rarely see advantageous mutations; they are almost always deleterious. Most detectable mutations, minus human medical intervention, are quite likely fatal. Some adaptations through mutation may, in fact, improve a species minimally, and may explain some things about that species, but they cannot lead to complete explanations of molecular life, the cell, integration at organ level, integrated body, world integration, and celestial integration. The "false eyes" on a moth's wings, for example, ostensibly represent a beneficial mutation for most observers, but a precursor, a rudimentary spot or weak color, would initially be more handicap than advantage.[41] Simple mutation and gradual evolution cannot account for the current existence of this moth.

Not insignificantly, Darwin's theory of natural selection actually falls in upon itself at this point when one considers that a system must have "minimal function" and must be capable of replication for natural selection in the environment to act upon it.[42] Darwinism ignores this one fundamental principle: *"Natural selection does not exist in prebiological molecules."*[43] It appears that scientists and lay people alike need to be reminded by codiscoverer of the DNA molecule, Francis Crick, that "the origin of life appears...to be almost a miracle, so many are the conditions that would have to be satisfied to get it going."[44]

Just as Darwinism is shown to be mathematically impossible, it is also theoretically baseless. In biology, the study of life, there are no laws "that permit the inference of all past and future states of the system...Biological systems tend not to be sufficiently closed to permit the formulation of such laws."[45] The eternal temptation is to find an all-encompassing formulation, a grand unified theory, but such reductionism is not appropriate or possible.[46]

Thus, like historical materialism and its demise as an explanatory and predictive model, evolutionary theory could never encompass as it sought to, and it is now spent. Darwin's dogma, founded in its inexact ideas, proselytized by scientific institutions and scientists for over a century, has misled many people and mocked the poetry of our existence.

Darwinism's danger

In opening Darwin's "black box," modern biology and medicine have opened a great paradox of positive potential and of possible peril. New exhilarating possibilities carry incumbent and heavy responsibilities. The misuse of knowledge and the miscarriage of these responsibilities may exact grave and dire consequences. The burden is upon us to remain ethically sensitive, morally upright, and willing to humbly relinquish preeminence to God. The pride of humanity, of scientists, could all too easily lead us down the path of self-destruction.

With the encouragement of and under the pedagogy of the scientific community, the primacy of God has been superceded by a belief in the near omnipotence of science and of human knowledge. Stretching our hands toward the "tree of life," seeking, in some manner, not only to displace the Creator but to *become* creator is more than just reckless. In our refusal to meet God at the foundation of existence, our spiritual cataracts will have us stumble to calamity in a wanton quest for mastery. We must realize His redemption, His omniscience, and remain vigilant that we do not lose sight of the greater, deeper truth.

This truth is that life was created by God. We may be its stewards; humanity may be able to modify aspects of procreation and creation, but we can never create a cell; we can never create life. The Huguenots taught us that life is within the Word. "Call to Me, and I will answer you, and show you great and mighty things, which you do not know" (Jer. 33:3). Under His will, He will share secrets with the hungry, inquiring human mind—a mind that must be based in humility rather than pride. Disaster might be imminent unless we seek the truth—the truth of God.

How wonderful it would be if the explosions of light within this restless, searching human mind could be centered in the background of the knowledge of the details of God's design that assert His artistry and supremacy. We must decide to overlook superficial red herrings and dive deeper to comprehend our part in the dance, our role as stewards of the astonishing and thrilling majesty of creation from the smallest cell to the outer galaxies to the vibrant, noisy rooms of human imagination. We have so little understanding of the whole that we have no choice but to seek God's Word and wisdom for more profound integration of our mere particles of knowledge. The essential attitude to life should be one of awe, appreciation, and respect while devoid of pride and a desire to transcend or manipulate without discipline.

Prejudices inspired by Darwin are not easily surrendered. The secular environment teaches us that "man is the measure," that we own ourselves and all that is in this world. Yet, when we accept God's universe, life, and us as His domain, His charge, and all designed for His glory, our eyes are truly opened to His handiwork. Then, with the veil lifted, with joyful and appreciative hearts, we behold an indescribable beauty that is the utter complexity and exactness of creation, and we are abundantly aware of the glorious mind of the Creator.

James P. Gills

James P. Gills, M.D. is founder and director of St. Luke's Cataract and Laser Institute in Tarpon Springs, Florida. Internationally respected as a cataract surgeon, Dr. Gills has performed more cataract extractions with lens implantations than anyone else in the world. He has pioneered many advancements in the field of ophthalmology to make cataract surgery safer and easier.

As a world-renown ophthalmologist, Dr. Gills has received innumerable medical and educational awards, highlighted by 1994–1999 listings in *The Best Doctors in America.* Dr. Gills is a clinical professor of ophthalmology at the University of South Florida and was named one of the Best Ophthalmologists in America in 1996 by ophthalmic academic leaders nationwide. He serves on the Board of Directors of the American College of Eye Surgeons, the Board of Visitors at Duke University Medical Center, and the Advisory Board of Wilmer Ophthalmological Institute at Johns Hopkins University. He has published more than 140 medical papers and authored eight medical textbooks. Listed in Marquis' *Who's Who in America,* Dr. Gills was Entrepreneur of the Year 1990 for the State of Florida and received the Tampa Bay Business Hall of Fame Award in 1993 and the Tampa Bay Ethics Award from the University of Tampa in 1995. In 1996 he was awarded the prestigious Innovators Award by his colleagues in the American Society of Cataract and Refractive Surgeons. In 2000 he was presented with the Florida Enterprise Medal by the Merchants Association of Florida, named

Humanitarian of the Year by the Golda Meir/Kent Jewish Center in Clearwater, and the Free Enterpriser of the Year by the Florida Council on Economic Education. In 2001 The Salvation Army presented to Dr. Gills its prestigious "Others" Award in honor of his lifelong commitment to service and caring.

Dr. Gills has dedicated his life to restoring much more than physical vision. He seeks to encourage and comfort the patients who come to St. Luke's. It was through sharing his insights with patients that he initially began writing on Christian topics. An avid student of the Bible for many years, he now has authored fourteen books dealing with Christian principles as well as physical fitness.

As an ultra-distance athlete, Dr. Gills has participated in forty-six marathons, including eighteen Boston marathons, and fourteen 100-mile mountain runs. In addition, he has completed five Ironman Triathlons in Hawaii and six Double Iron Triathlons. Dr. Gills has served on the national Board of Directors of the Fellowship of Christian Athletes and in 1991 was the first recipient of their Tom Landry Award.

Married in 1962, Dr. Gills and his wife, Heather, have raised two children, Shea and Pit. Shea Gills Grundy, a former attorney now full-time mom, is a graduate of Vanderbilt University and Emory University Law School. She and husband, Shane Grundy, M.D., presented the Gills with their first grandchildren—twins, Maggie and Braddock, and three years later a third child, James Gills Grundy. The Gills' son, J. Pit Gills, M.D., ophthalmologist, received his medical degree from Duke University Medical Center and in 2001 joined the St. Luke's staff. "Dr. Pit" is married to Joy Parker-Gills. They are the proud parents of James Pitzer Gills IV.

Intelligent agents have unique causal powers that nature does not. When we observe effects that we know only agents can produce, we rightly infer the presence of a prior intelligence even if we did not observe the action of the particular agent responsible. Since DNA displays an effect (namely, information content) that in our experience only agents can produce, intelligent design (and not apparent design) stands as the best explanation for the information content in DNA.

—Stephen C. Meyer, from "Word Games" in *Signs of Intelligence*

FOUR

Chapter 4

DNA, Design, and the Origin of Life

By Charles B. Thaxton, Ph.D, President, KONOS Connection

The classical design argument looked at order in the world and concluded that God must have caused it. Archdeacon William Paley in the nineteenth century refined the argument.[1] He also gave it perhaps its most eloquent and persuasive formulation. Paley looked at the order of human artifacts and compared it to the order in living beings. If human intelligence was responsible for artifacts, reasoned Paley, then some intelligent power greater than man must have accounted for living beings.

The major problem with this design argument was its claim to reason from order in the world to a supernatural designer. Paley did not provide any uniform experience of the supernatural, which alone could make good his claim. As valid as this objection was, however, only philosophers seemed concerned about it. It was an argument by Charles Darwin that raised doubt for most people concerning true design in the world.[2] According to Darwin, natural selection produced *apparent* design, which the faithful mistook for true design. So the matter has stood in the scientific community and the world at large for a century.

Scientific discoveries made in this century, however, threaten to change the outlook fundamentally in regards to design. However, few outside the relevant disciplines seem aware of it. I am referring to developments in relativity theory and quantum mechanics,[3] neurophysiology,[4] information theory,[5] and molecular biology, particularly the elucidation of the structure of DNA (deoxyribonucleic acid).[6] I shall focus my remarks on DNA and its relation to design and the origin of life.

Due to advances in molecular biology, the process of reproduction, or *self-replication,* has become better understood. At the core of this process is the DNA molecule. Though not itself alive, DNA is usually regarded as the *sine qua non* of life. DNA is considered the identifying mark of a living system. We judge something as living if it contains DNA.

Molecular biology has shown us how extremely intricate living things are, especially the genetic code and the genetic process. Interestingly enough, the genetic code can be best understood as an analogue to human language. It functions exactly like a code. Indeed, it is a code—it is a molecular communication system within the cell.

A sequence of chemical *letters* stores and transmits the communication in the cell. Communication is possible no matter what symbols are used as an alphabet. The twenty-six letters we use in English, the thirty-two Cyrillic letters used in the Russian language, or the four-letter genetic alphabet—all serve in communication.

In recent years, scientists have applied information theory to biology, and in particular to the genetic code. *Information theory* is the science of message transmission developed by Claude Shannon and other engineers at Bell Telephone Laboratories in the late 1940s. It provides a mathematical means of measuring information. Information theory applies to any symbol system, regardless of the elements of that system. The so-called Shannon information laws apply equally well to human language, Morse code, and the genetic code.

The conclusion drawn from the application of information theory to biology is that there exists a structural identity between the DNA code and a written language. H. P. Yockey notes in the *Journal of Theoretical Biology:*

> It is important to understand that we are not reasoning by analogy. The sequence hypothesis [that the exact order of symbols records the information] applies directly to the protein and the genetic text as well as to written language and therefore the treatment is mathematically identical.[7]

This development is highly significant for the modern origin of life discussion. Molecular biology has now uncovered an analogy between DNA and written human languages. It is more than an analogy; in fact, in terms of structure, the two are "mathematically identical." In the case of written messages, we have uniform experience that they have an intelligent cause. What is *uniform experience?* It simply means that people everywhere observe a certain type of event always in association with a certain type of cause. When we find evidence that a similar event happened in the past, it is reasonable to infer it had a similar cause. As I shall argue, based on uniform experience there is good reason to accept an intelligent cause for the origin of life as well.[8, 9]

Two Kinds of Order

You may recognize this argument for an intelligent cause of life. It is a form of the design argument that has been popular among theists for centuries. The design argument makes use of the same mode of reasoning used in the historical sciences today—namely, the argument from analogy. The design argument assumes that the order we see in the world around us bears an analogy to the kind of order exhibited by human artifacts, by tools and machines, and by works of art. Since the two kinds of order are similar, the cause of one must be similar to the cause of the other. The order in human artifacts is the result of human intelligence. Therefore, the order in the world must be the result of an intelligent being we call the *Creator.*

The argument from molecular biology is a modern restatement of the argument from design, with a few significant refinements. The older design argument went straight from order in the universe to the existence of God. From time immemorial, the beauty of the birds and flowers, the cycle of the seasons, the remarkable adaptations in animals, have led people to posit some type of intelligent cause behind it all. Not just Christians, but a wide range of believers in some form of intelligence have buttressed their belief by appealing to the wonderful order and complexity in the world.[10, 11]

During the Scientific Revolution of the seventeenth century, the argument from order took on even greater force. Scientists studied the intricate structures in nature in a depth and detail unknown in previous ages. Many became more convinced than ever that such order required an intelligent cause. Isaac Newton expressed a common sentiment when he declared, "This most beautiful system of the sun, planets, and comets could only proceed from the counsel and dominion of an intelligent and powerful Being."[12]

The argument from design has always been the argument most widely accepted by scientists. It is the most empirical of the arguments for God based as it is on observational premises about the kind of order we discover in nature.[13] Ironically, it was also the Scientific Revolution that eventually led many to reject the argument from design. Repeatedly, scientists discovered natural causes for events that until then had been mysterious. If natural causes could explain these things, perhaps they could explain everything else, too. Do we really need an intelligent cause to explain the order of the world?

Take, for example, the structure of a snowflake. The intricate beauty of a snowflake has led many a believer to exclaim upon the wisdom of the Creator. Yet the snowflake's structure is nothing mysterious or supernatural. It is explained by the natural laws that govern the crystallization of water as it freezes. The argument from design claims that the order we see around us cannot have arisen by natural causes. The snowflake seems to refute that claim. It demonstrates that at least some kinds of order can arise by natural causes. And if matter alone can give rise to order in some instances, why not in all others as well? Why do we need to appeal to an intelligent being any more to explain the origin of the world? We need only continue to search for natural causes. Many materialists today use this argument.[14]

What is coming to light through the application of information theory is that there are actually two kinds of order. The first kind (the snowflake's) arises from constraints within the material the thing is made of (the water molecules). We cannot infer an intelligent cause from it, except possibly in the remote sense of something behind the natural cause. The second kind, however, is not a result of anything within matter itself. It is in principle opposed to anything we see forming naturally. This kind of order does provide evidence for an intelligent cause.

The Difference It Makes

Let's explain these two kinds of order in greater detail. As you travel through various parts of the United States, you may come across unusual rock formations. If you consult a tourist's guide, you will learn that such shapes result when more than one type of rock make up the formation. Because their mineral composition varies, some rocks are softer than others. Rain and wind erode the soft parts of the formation faster than the hard parts, leaving the harder sections protruding. In this way, the formation may take on an unlikely shape. It may even come to resemble a familiar object like a face.

In other words, the formation may look as though it were deliberately carved. However, on closer inspection, say from a different angle, you notice the resemblance is only superficial. The shape invariably accords with what erosion can do, acting on the natural qualities of the rock (soft parts worn away, hard parts protruding). You therefore conclude the rock formed naturally. Natural forces suffice to account for the shape you see.

Now let's illustrate a different kind of order. Say in your travels that you visit Mount Rushmore. Here you find four faces on a granite cliff. These faces do not follow the natural composition of the rock: The chip marks cut across both hard and soft sections.[15] These shapes do not resemble anything you have seen resulting from erosion. In this case the shape of the rock is not the result of natural processes. Rather, you infer from uniform experience that an artisan has been at work. The four faces were intelligently imposed onto the material.

None of us find it difficult to distinguish between these two kinds of order—the one produced naturally and the other by intelligence. To come back to the argument from design, the question is: Which kind of order do we find in nature?

If we find only the first kind, then our conclusion will be that natural causes suffice to explain the universe as we see it today. An intelligent cause, if there is one, is merely a distant First Cause. It is a deistic kind of God who created matter with certain tendencies and then stood back to let these work themselves out mechanically.

If, on the other hand, we find any instances of the second kind of order, the kind produced by intelligence, these will be evidence of the activity of an intelligent cause. Science itself would then point beyond the physical world to its origin in an intelligent source.

It is easy enough to find examples of the first kind of order. The snowflake was one. The properties of the atoms that compose a snowflake determine its crystalline structure. Wind and temperature explain cloud shapes. Ripples of sand on a beach result from the impact of wind and waves. The waves of the sea form by wind, gravity, and the fluid properties of water. None of these go beyond what we expect to result naturally, given the properties of the material itself. The beauty of a sunset may inspire poets, but natural causes suffice to explain it.

The most pervasive example of the second kind of order is life itself.

A Code in Miniature

One of the greatest scientific developments of the twentieth century has been the discovery of the DNA code. DNA is the famous molecule of heredity. Each of us begins as a tiny ball about the size of a period at the end of a sentence. All our physical characteristics—height, hair color, eye color, and so forth—are *spelled out* in our DNA. It guides our development into adulthood.

The DNA code is quite simple in its basic structure (although enormously complex in its functioning). By now most people are familiar with the double helix structure of the DNA molecule. (See page 44.) It is like a long ladder, twisted into a spiral. Sugar and phosphate molecules form the sides of the ladder. Four bases make up its *rungs*. These are adenine, thymine, guanine, and cytosine. These bases act as the *letters* of a genetic alphabet. They combine in various sequences to form words, sentences, and paragraphs. These base sequences are all the instructions needed to guide the functioning of the cell.

The DNA code is a genetic *language* that communicates information to the cell. The cell is very complicated, using many DNA instructions to control its every function. The amount of information in the DNA of even the single-celled bacterium *E. coli* is vast indeed. It is greater than the information contained in all the books in any of the world's largest libraries. The DNA molecule is exquisitely complex and extremely precise—the *letters* must be in a very exact sequence. If they are out of order, it is like a typing error in a message. The instructions that it gives the cell are garbled. This is what a mutation is.

The discovery of the DNA code gives the argument from design a new twist. Since life is, at its core, a chemical code, the origin of life is the origin of a code. A code is a very special kind of order. It represents "specified complexity."[16] To understand this term, we need to take a brief excursion into information theory as it applies to biology.

Measuring Information

"One if by land, two if by sea." Paul Revere did not know information theory, but he was using its principles correctly. A simple but effective code informed the patriots of the British route of approach.

Information theory realizes an important goal of mathematicians: to make information measurable. It finds its place in biology through its ability to measure organization and to express it in numbers. Biologists have long recognized the importance of the concept of organization. However, little practical was possible until there was a way to measure it. Organization stated in terms of information does this. "Roughly speaking," says Leslie Orgel, "the information content of a structure is the minimum number of instructions needed to specify the structure."[17] The more complex a structure is, the more instructions needed to specify it. Random structures require very few instructions at all. If you want to write out a series of nonsense letters, for example, here is all you do. The only instructions necessary are, "Write a letter between A and Z," followed by "now do it again," ad infinitum.

A highly ordered structure likewise requires few instructions if its order is the result of a constantly repeating structure. A whole book filled only with the sentence "I love you" repeated over and over is a highly ordered series of letters. A few instructions specify which letters to choose and in what sequence. These instructions followed by "now do it again" as many times as necessary completes the book. By contrast with either random or ordered structures, complex structures require many instructions. If we wanted a computer to write out a poem, for example, we would have to specify each letter. That is, the poem has a high information content.

Specifying a Sequence

Information in this context means the precise determination, or specification, of a sequence of letters. We said above that a code represents "specified complexity." We are now able to understand what *specified* means: The fewer choices there are about fulfilling each instruction, the more highly specified a thing is.

In a random situation, options are unlimited, and each option is equally probable. In generating a list of random letters, for instance, there are no constraints on the choice of letters at each step. The letters are unspecified.

An ordered structure, on the other hand, like our book of "I love yous," is highly specified—that is, redundant. Each letter is specified. Nonetheless, it has a low information content, as noted before, because the instructions needed to specify it are few. Ordered structures and random structures are similar in that both have a low information content. However, they differ in that ordered structures are highly specified.

A complex structure like a poem is likewise highly specified. It differs from an ordered structure, however, in that it is not only highly specified, but also has a high information content. Writing a poem requires new instructions to specify each letter.

To sum up, information theory has given us tools to distinguish between the two kinds of order we spoke about at the beginning.

Lack of order—randomness—is neither specified nor high in information.

The first kind of order is the kind found in a snowflake. Using the terms of information theory, a snowflake is specified but has a low information content. Its order arises from a single structure repeated over and over. It is like the book filled with "I love you." The second kind of order, the kind found in the faces on Mount Rushmore, is both specified and high in information.

Life Is Information

Molecules characterized by specified complexity make up living things. These molecules are, most notably, DNA and protein. By contrast, nonliving things fall into one of two categories. Either they are unspecified and random (like lumps of granite and mixtures of random nucleotides), or they are specified but simple (like snowflakes and crystals). A crystal fails to qualify as living because it lacks complexity. A chain of random nucleotides fails to qualify because it lacks specificity.[18] No nonliving things (except DNA and protein in living things, human artifacts, and written language) have specified complexity.

For a long time biologists overlooked the distinction between two kinds of order (simple, periodic order vs. specified complexity). Only recently have they appreciated that the distinguishing feature of living systems is not *order* but *specified complexity.*[19] The sequence of nucleotides in DNA, or of amino acids in a protein, is not a repetitive order like a crystal. Instead it is like the letters in a written message. A message is not composed of a sequence of letters repeated over and over. It is not, in other words, the first kind of order.

Indeed, the letters that make up a message are in a sense random. There is nothing inherent in the letters *g-i-f-t*, which tells us the word means "present." In fact, in German the same sequence of letters means "poison." In French the series is meaningless. If you came across a series of letters written in the Greek alphabet and didn't know Greek, you wouldn't be able to read it. Nor would you be able to tell if the letters formed Greek words or were just groupings of random letters. There is no detectable difference.

What distinguishes a language is that certain random groupings of letters have come to symbolize meanings according to a given symbol convention. Nothing distinguishes the sequence *a-n-d* from *n-a-d* or *n-d-a* for a person who doesn't know any English. Within the English language, however, the sequence *a-n-d* is very specific and carries a particular meaning.[20]

There is no detectable difference between the sequence of nucleotides in *E. coli* DNA and a random sequence of nucleotides.[21] Yet within the *E. coli* cells, the sequence of "letters" of its DNA is very specific. Only that particular sequence is capable of biological function.

The discovery that life in its essence is information inscribed on DNA has greatly narrowed the question of life's origin. It has become the question of the origin of information. We now know there is no connection at all between the origin of order and the origin of specified complexity. There is no connection between orderly repeating patterns and the specified complexity in protein and DNA. We cannot draw an analogy, as many do, between the formation of a crystal and the origin of life. We cannot argue that since natural forces account for the crystal, then they account for the structure of living things. The order we find in crystals and snowflakes is not analogous to the specified complexity we find in living things.[22]

Are we not back to a more sophisticated form of the argument for design? With the insights from information theory, we need no longer argue from order in a general sense. Order with low information content (the first kind) does arise by natural processes. However, there is no convincing experimental evidence that order with high information content (the second kind or specified complexity) can arise by natural processes. Indeed, the only evidence we have in the present is that it takes intelligence to produce the second kind of order.

The Present As the Key to the Past

Scientists can synthesize proteins suitable for life. Research chemists produce things like insulin for medical purposes in great quantities. The question is, *how* do they do it? Certainly not by simulating chance or natural causes. Only by highly constraining the experiment can chemists produce proteins like those found in living things. Placing constraints on the experiment limits the choices at each step of the way. That is, it adds information. If we want to speculate on how the first informational molecules came into being, the most reasonable speculation is there was some form of intelligence around at the time.

The scientists searching for extraterrestrial intelligence (ETI) would recognize the kind of order inherent in a decodable signal from space as evidence of an intelligent source.[23] These scientists have never seen an extraterrestrial creature. However, they would recognize the similarity of a message from space to messages generated by human intelligence. In the same way, we note that the structure of protein and of DNA has a high information content.[24] We recognize its similarity to information (like poems and computer programs) generated by human intelligence. Therefore we may properly infer that the source of information on the molecular level was likewise an intelligent being.[25] Furthermore, we know of no other source of information. Efforts to produce information-bearing molecules by chance or natural forces have failed.[26] We have not seen the Creator, nor observed the act of creation. However, we recognize the kind of order that only comes from an intelligent being.

With the new data from molecular biology and information theory, we can now argue for an intelligent cause of the origin of life. It is based on the analogy between the DNA code and a written message. We cannot identify that source any further from the scientific data alone. We cannot supply a name for that intelligent cause. We cannot be sure from the empirical data on DNA whether the intelligence is within the cosmos, but off the earth, as asserted by Hoyle and Wickramasinghe.[27] It might be beyond the cosmos, as historic theism maintains.

All we can say is that given the structure of a DNA molecule, it is certainly legitimate to conclude that an intelligent agent made it. Life came from a *who* rather than a *what.* We may be able to identify that agent in greater detail by other arguments. We may, for example, gain insight from historical, philosophical, or theological argument, or by considering the relevant lines of evidence from other areas of science. However, from scientific data on DNA alone we can argue only to an intelligent cause.

Let's spell out the steps of the argument more explicitly. Does it in fact satisfy the principle of analogy? Yes, it does. First, we establish that an analogy does exist between the kind of order we see in living things and the kind we see in some other phenomena made by human intelligence.[28] We have an abundance of examples of specified complexity—books, machines, bridges, works of art, computers. All these are human artifacts. In our experience only human language and human artifacts match the specified complexity exhibited by protein and DNA. Second, we ask what is the source of the order in these modern examples? We know by uniform experience that its source is human intelligence.

The only remaining question is whether it is legitimate to use this reasoning to infer the existence of an intelligent cause *before* the existence of human beings. I would argue it is. A phenomenon from the past, known by uniform experience to be like that caused only by an intelligent source, is itself evidence that such a source existed. Even the simplest forms of life, with their store of DNA, are characterized by specified complexity. Therefore life itself is *prima facie* evidence that some form of intelligence was in existence at the time of its origin.[29]

It is true that our actual experiential knowledge of intelligence is limited to carbon-based organisms, particularly human beings. However, scientists already speculate on some other kinds of intelligence—that is, nonhuman—when they seriously seek to discover ETIs.[30] Some even argue that intelligence exists in complex nonbiological computer circuitry. Scientists today conceive of intelligence freed from biology as we know it. Then why can we not conceive of an intelligent being existing before the appearance of biological life in this planet?

Uniform Experience

In scientific terms, the analogy criterion is the same thing as the principle of uniformity. It is the dictum that our theories of the past must invoke causes similar to those acting in the present. David Hume was getting at the same idea with his phrase "uniform experience."[31]

As regards the origin of life, our uniform experience is that it takes an intelligent agent to generate information, codes, messages. As a result, it is reasonable to infer there was an intelligent cause of the original DNA code. DNA and written language both exhibit the property of specified complexity. Since we know an intelligent cause produces written language, it is legitimate to posit an intelligent cause as the source of DNA.

We have now defined the DNA code as a message. It is now clear that the claim that DNA arose by material forces is to say that information can arise by material forces. However, the material base of a message is completely independent of the information transmitted. The material base could not have anything to do with the message's origin. The message transcends chemistry and physics.[32]

When I say that a message is independent of the medium that conveys it, I mean that the materials used to send a message have no affect whatever on the content of the message. The content of "apples are sweet" does not change when I write it in crayon instead of ink. It is unaffected by a switch to chalk or pencil. I can say the same thing if I use my finger and write it in the sand. I can also use smoke and write it in the sky. I can translate it into the dots and dashes of Morse code. Even people holding up posters at a baseball game can transmit the same information.

The point is, there is no relationship at all between the information and the material base used to transmit it. The ink or chalk I use to write "apples are sweet" does not itself look red, nor does it taste sweet like an apple. There is nothing in the ink molecules that compels me to write precisely or only that particular sentence. The information transmitted by my writing is not within the ink I use to write it. Instead, an outside source imposes information upon the ink using the elements of a particular linguistic symbol system.

The information within the genetic code is likewise entirely independent of the chemical makeup of the DNA molecule. The information transmitted by the sequence of bases has nothing to do with the bases themselves. There is nothing in the chemicals themselves that originates the communication transmitted to the cell by the DNA molecule.

These rather obvious facts are devastating to any theory that assumes life first arose by natural forces. Such theories dominate the intellectual landscape today. Some theories assume that energy flow organized the chemicals to produce information in the first DNA molecule.[33, 34, 35] Others assume that self-organizing properties within the chemicals themselves created the information.[36] Yet this is tantamount to saying the material used to transmit information also produced it. It is as though I were to say it was the chemical properties of the ink itself that caused me to write, "Apples are sweet."

We can state our case even more strongly. To accept a material cause for the origin of life actually runs counter to the principle of uniformity. Uniform experience reveals that only an intelligent cause regularly produces specified complexity. To be sure, we may still posit a nonintelligent, material cause as the source of specified complexity, even though we do not regularly observe it. We may argue that in the rare occurrence, in spite of its trivially small probability, such an event might happen. The problem is, however, that to argue this way is no longer to do science. Regular experience, not negligible probabilities and remote possibilities, is the basis of science.

Darwin convinced many of the leading intellectuals in his time that design in the world is only apparent, that it is the result of natural causes. Now, however, the situation has taken a dramatic turn, though few have recognized its significance. The elucidation of DNA and the unraveling of the secrets of the genetic code have opened again the possibility of seeing true design in the universe.*

*"DNA, Design, and the Origin of Life" was published, in a slightly different version, in *Cosmic Pursuit*, March 1998. It is reprinted here by permission of the author, Dr. Charles Thaxton.

Charles Thaxton

Charles Thaxton, who has taught for many years in the Department of Natural Sciences at Charles University in Prague, Czech Republic, is one of the leading architects of the Intelligent Design movement. After receiving his Ph.D. in chemistry from Iowa State University, he lectured at the L'Abri Study Center in Switzerland and held postdoctoral fellowships in the history of science at Harvard University and in molecular biology at Brandeis University.

Charles Thaxton's groundbreaking work *The Mystery of Life's Origin*, coauthored with Walter Bradley and Roger Olsen, is a penetrating, comprehensive critique of the field that studies "chemical evolution"—the origin of the first living cells. It received wide praise, even among scientists who toil in this area. Most notably, it helped to spark a growing skepticism about the "prebiotic soup" hypothesis of the origin of life. *The Soul of Science*, written by Thaxton with Nancy Pearcey, traces the crucial (and often unnoticed) role of *philosophy* in shaping science—from Aristotle and the medieval alchemists all the way to relativity, quantum physics, and modern molecular biology.

The fascinating chapter you just read stands as one of the clearest and most compelling presentations of the case for an intelligent origin of the vast information content of the DNA molecule. This spiral-ladder thread contains a unique message, which is coded inside every living thing.

For I am well aware that scarcely a single point is discussed in this volume on which facts cannot be adduced, often apparently leading to conclusions directly opposite to those at which I have arrived. A fair result can be obtained only by fully stating and balancing the facts and arguments on both sides of each question...

—Charles Darwin from "Introduction" in *The Origin of Species*

FIVE

Chapter 5

Darwinism on Trial

BY PHILLIP JOHNSON, J.D.

When Phillip Johnson began to question the credibility of Darwinism in 1987, he launched his own research on the topic as an "outside auditor," a respectful critic specializing in the rhetorical techniques that are employed in public argumentation. Johnson's voluminous speaking and writing on origins over the past decade have thus focused on how language is used by evolutionists to "persuade" the public. He has powerfully educated his audiences on several bedrock issues in this area—especially on *how evidence is used (or ignored) by Darwinists, and how to spot the sleight-of-hand when assumptions are slipped into an argument in order to protect a weak case.*

Johnson has penned six books on Darwinism and its philosophical bedfellow, naturalism. His first book, *Darwin on Trial*, is a wide-ranging critique that began as a yearlong research project in England. He had been prompted to write by reading a vigorously pro-Darwinist book, *The Blind Watchmaker* by Oxford biologist Richard Dawkins, and by Denton's skeptical treatment, *Evolution: A Theory in Crisis.* Johnson elaborated his critique through three more years of intense research into the theory's evidential foundations. The deeper that Johnson dug into the scientific literature, the more he was convinced that the evidence for naturalistic large-scale evolution (or "macroevolution") was "somewhere between weak and nonexistent."[2]

He concluded in *Darwin on Trial* that the crucial underpinnings of Darwinism are not evidential, but philosophical. Simply stated, Darwinists assume that the universe is a "closed system" of causes and effects, which cannot be influenced by anything outside, like a creator. This philosophy goes by various labels, especially "scientific materialism" and "naturalism."

Darwin on Trial was published in revised edition in 1993 to respond to a variety of criticisms brought against the book, including an acidic attack published by Stephen Jay Gould. Since then, Johnson has penned five other books on the related topics of Darwinism, science, naturalism, and modern culture. All along his path of research and writing, Johnson has been careful to submit his work to the critical scrutiny of leading Darwinists. Among these were three preeminent experts on evolution—William Provine of Cornell (who began inviting Johnson to speak in his class in Ithaca, New York), philosopher Michael Ruse of Florida State University (who agreed to enter dialogue with Johnson at a major conference on evolution), and British paleontologist Colin Patterson (who assisted Johnson by checking for errors in his first book draft in England). Johnson even convened a seminar of fellow professors at Berkeley, many of them in the life sciences and philosophy of science, to review his work.

One of the key occasions where Johnson met with scholars to present his critique was at a December 1989 meeting of a dozen academicians at the Campion Center, a Jesuit conference center near Boston. Attending this weekend seminar-retreat were scholars from fields as diverse as the history of astronomy (Owen Gingerich), theology and science (Langdon Gilkey), and origin-of-life chemistry (Charles Thaxton). Most were believers in Darwinian evolution, although a few, such as Thaxton, were skeptical.

Before long, Johnson's work became the center of attention, and the "Campion Summary" paper, from which this chapter has been adapted, was Johnson's attempt to help the participants by distilling his lengthy paper (which had also been distributed to the attendees) into its key points.

What was most remarkable about this weekend was the outbreak of a verbal duel between Johnson and the late renowned evolutionist Stephen Jay Gould. This interchange erupted shortly after the accuracy of Johnson's critique was certified by a top paleontologist who was attending the meeting from the University of Chicago. Fellow paleontologist Gould, who joined the seminar on Saturday morning, could not let such tacit approval stand and promptly launched a salvo of vehement criticisms of Johnson's work. Within minutes, the room was mesmerized as it witnessed a fast and furious debate between Johnson and Gould, which lasted for half an hour. (It may be helpful here to conjure images of the extended clashing of "light sabers" in a Star Wars movie.) Many who attended on both sides of the issue later described this amazing exchange as a "draw."

Johnson was very satisfied with the entire meeting and considered the interaction with Gould a significant step in the development of his critique of Darwinism. Simply put, he had held his own with one of the theory's chief defenders of modern times.

We hope that as you read this adaptation of Johnson's summary paper from the Campion meeting, you will not only sense the clarity and cogency of his criticisms, but will also realize how timeless Johnson's words were. Essentially all of the major points of the empirical and philosophical critique of Darwinism, seen today in the writings of Intelligent Design theorists, are seen here in seed form. Enjoy!

—EDITOR'S NOTE

The Campion Summary Paper

The important issue is not the relationship of science and creationism, but the relationship of science and materialist philosophy.

Creationism to most people means biblical literalism, and specifically the doctrine that all "basic kinds" of living organisms were separately created by God within the space of a single week about six thousand years ago. I have no interest in promoting or even discussing creationism in that sense.

The question I raise is not whether science should be forced to share the stage with some biblically based rival known as *Creationism*, but whether we ought to be distinguishing between the doctrines of scientific materialist philosophy and the conclusions that can legitimately be drawn from the empirical research methods employed in the natural sciences.

Scientific materialism (or naturalism, as in my 1988 draft) is the philosophical doctrine that everything real has a material basis, that the path to objective knowledge (as distinguished from subjective belief) is exclusively through the methods of investigation accepted by the natural sciences, and that teleological conceptions of nature ("we are here for a purpose") are invalid. To a scientific materialist there can be no "ghost in the machine," no nonmaterial intelligence that created the first life or guided its development into complex form, and no reality that is in principle inaccessible to scientific investigation, such as the supernatural.

The metaphysical assumptions of scientific materialism are not themselves established by scientific investigation, but rather are held *a priori* as unchallengeable and usually unexamined components of the "scientific" worldview. Materialist science therefore does not investigate *whether* the first living organisms evolved from nonliving chemicals without the intervention of any preexisting intelligence; likewise, it does not investigate *whether* the emergence of complex plants and animals, human consciousness, and so on was the product of purely natural (mindless, nonteleological) processes. The naturalistic evolution of life from prebiotic chemicals, and its subsequent naturalistic evolution into complexity and humanity, is *assumed* as a matter of first principle, and the only question open to investigation is how this naturalistic process occurred.

The question is whether this refusal to consider any but naturalistic explanations has led to distortions in the interpretation of empirical evidence, and especially to claims of knowledge with respect to matters about which natural science is in fact profoundly ignorant.

The Dominance of Neo-Darwinism

The continued dominance of neo-Darwinism is the most important example of distortion and overconfidence resulting from the influence of scientific materialist philosophy upon the interpretation of the empirical evidence.

Claims that natural selection is a force of stupendous creative power that is capable of crafting the immensely complex biological structures that living creatures possess in such abundance are not supported by experimental evidence or observation. The analogy to artificial selection (where conscious intelligence strives to produce greater variety) is faulty. In any case, artificial selection does not continue to produce change in a particular direction indefinitely. Observational evidence (such as the famous peppered moth study) shows mainly cyclical changes in the relative frequency of characteristics already present in the population. There is circumstantial evidence pointing to somewhat more impressive changes (such as circumpolar gulls or Hawaiian fruit fly species), but the empirical evidence gives no reason for confidence that natural selection has the creative power, regardless of the amount of time available, to build up complex organs from scratch or to change one body plan into another.

Natural Selection

Darwinists deny that natural selection is a tautology when the issue surfaces explicitly. The concept is not inherently tautological, and it is capable of being stated in testable form. It is also capable of being formulated and used as a tautology, however, and in practice Darwinists continually employ it as the invisible cause of whatever change or lack of change seems to have occurred. If new life forms appear, it is due to creative natural selection; if old forms fail to change, the credit goes to stabilizing selection; and the survival of some groups during mass extinctions is explained by their greater "resistance to extinction."

The claim that selection in combination with random micro-mutations can craft new forms and complex organs by gradual steps is disconfirmed by the impossibility of proposing plausible advantageous intermediate forms in many cases. This difficulty can be met with various ad hoc speculations, such as hypothetical mutations in rate genes affecting embryonic development, but experimental confirmation that such processes can create complex organs and new body plans is unavailable. Whether the Darwinistic evolution of wings, eyes and so on is conceivably possible is not the question here. The question is whether it is more than a speculative possibility.

The decisive disconfirmation of neo-Darwinism comes from the fossil record. Even if we generously grant the assumption that neo-Darwinist macroevolution is capable of producing basic changes, it does not appear to have done so.

Darwin's hypothesis requires the existence of an immense

quantity of transitional forms that became extinct as they were gradually replaced by better-adapted descendants. That the fossil record shows a consistent pattern of sudden appearance of new forms, followed by stasis—that is, the pervasive absence of the indispensable transitional intermediates—was a problem in 1859, but it is a far more serious problem when the condition persists after 130 years of determined search for the missing transitional intermediates.

The Cambrian explosion of the animal phyla in a world previously composed of algae and bacteria (excepting the Ediacara fauna, which do not fill the gap as transitional intermediates), and the failure of life to diversify into new phyla thereafter, can be reconciled with Darwinism only by the strenuous application of ad hoc hypotheses.[3]

The importance of mass extinctions—practically invisible for many decades due to the influence of Darwinist prejudice—further disconfirms Darwinist claims that continuous natural selection gradually weeds out the unfit and accounts for the absence of surviving transitional intermediates. Again, the question is not whether ad hoc hypotheses can be invented to save Darwinism in the teeth of all this unfavorable evidence. The question is whether the evidence, fairly considered in its entirety, gives us cause for confidence that Darwinist evolution took life all the way from the hypothetical first living microorganism to where we find it today.

The Influence of Scientific Materialist Philosophy

The refusal (or inability) of the scientific establishment to acknowledge that Darwinism is in serious evidential difficulties and probably false as a general theory is due to the influence of scientific materialist philosophy and certain arbitrary modes of thought that have become associated with the scientific method. Science requires a paradigm or organizing set of principles, and

Darwinism has fulfilled this function for more than a century. It is the grand organizing theoretical principle for biology—a statement that does not imply that it is true.

Once established as orthodox, a paradigm customarily is not discarded until it can be replaced with a new and better paradigm that is acceptable to the scientific community. Disconfirming evidence (anomalies) can always be classified as "unsolved problems," and the situation remains satisfactory for researchers because even an inadequate paradigm can generate an agenda for research.

To be acceptable, a paradigm must conform to the philosophical tenets of scientific materialism. For example, the hypothesis that biological complexity is the product of some preexisting creative intelligence or vital force is not acceptable to scientific materialists. They do not fairly consider this hypothesis and then reject it as contrary to the evidence; rather they disregard it as inherently ineligible for consideration.

Given the above premises, something very much like Darwinism simply must be accepted as a matter of logical deduction, regardless of the state of the evidence. Random mutation and natural selection must be credited with shaping biological complexity, because nothing else could have been available to do the job. Hence even those evolutionary biologists who are most frank in acknowledging that Darwinism is in trouble frequently end up saying in the next sentence that no reputable biologist seriously doubts the importance of (creative) natural selection in evolution. Because the escape from Darwinism seems to lead nowhere, Darwinism for scientific materialists is inescapable.

What makes this situation particularly misleading is a confusion (at times convenient for Darwinists) about how the category "science" relates to the category "truth." Outside critics who point out the disconfirming evidence are frequently put off with the rejoinder that they do not understand "how science works," or with the disclaimer that Darwinists are noticing that theistic or teleological interpretations "should not be taught in science class."

Having put aside certain important possibilities as inherently unworthy of consideration, however, Darwinists do not hesitate to assert that their conclusions are objectively true—that is, evolution (naturalistic evolution) is a fact: Natural selection really has the powers claimed for it. These statements carry the implication that the philosophical premises on which they are based are also objectively true and, therefore, that competing philosophical premises are false. To put this somewhat abstract point in the vernacular: If God was around and capable of creating, there is no reason to assume that those Cambrian phyla must have evolved through purely naturalistic mechanisms.

Unavoidable Difficulties

The difficulties of Darwinism cannot be avoided by retreating to some supposedly unchallengeable "fact" of evolution or by proposing "alternatives to phyletic gradualism" that attempt to occupy a middle ground between Darwinism and saltationism.

Some Darwinists have attempted to distinguish a purportedly indisputable "fact" of evolution from the concededly debatable theory of evolution by gradualistic natural selection. But what, precisely, is the fact to which they refer? The occurrence of microevolution is a fact, and that life has a common biochemical basis (the DNA code, etc.) is also a fact. The existence of natural relationships of greater and lesser similarity (classification) is also a fact. If *evolution* is merely a shorthand expression for *microevolution* and *relationship*, then its use tells us nothing about how those relationships came into being.

In practice, the "fact of evolution" turns out to be Darwinism. Thus Gould's essay "Evolution As Fact and Theory" distinguishes the fact of evolution from a hyper-Darwinism (pan-selectionism) that Darwin himself repudiated, not from Darwinism properly understood. This reservation is unavoidable because the important claim of "evolution" is not that fish and man have certain common features, but that it is possible for a fish ancestor to produce a descendant human being, given sufficient time and the right conditions, without intelligent intervention. Absent the supposed creative power of natural selection, the transformation of fish to man would be nearly as miraculous as the instantaneous creation of man from the dust of the earth.

Nonmaterialistic theories of evolution (such as that God or a life force takes an active supervisory role) are nearly as unacceptable to scientific materialists as outright creationism. Saltationism likewise is not really evolution at all, but a meaningless halfway point between creation and evolution. As Dawkins put it, "You can call the creation of man from the dust of the earth a saltation."[4]

Semi-saltationist "alternatives to phyletic gradualism" are not genuine alternatives to Darwinism. To the extent that "punctuationalists" are merely saying that Darwinist evolution occurs rapidly (in geological terms) and in populations so small as to escape preservation in the fossil record, they are playing a variation on the usual theme that the fossil record is incomplete. As such they assume the whole point at issue, which is whether Darwinist macroevolution actually happened. To the extent that the punctuationalists are saying that the missing intermediates never existed as living creatures, they incur the disadvantages that led to the discrediting of Goldschmidt. The former option seems to be the lesser evil, and so we are now being assured that punctuated equilibrium is within the Darwinist framework and really just an elaboration of the implications for paleontology of Ernst Mayr's theory of speciation.

Retreat to the fact of evolution therefore inevitably brings Darwinism in again through the back door, without the need for defending its vulnerable aspects. The person who accepts the fact of evolution, thinking that he is acknowledging only that living things are related, will quickly learn that he is deemed to have accepted also that the relationships are based upon common ancestry.[5] If fish and man are descended from a common ancestor, then the immense differences that distinguish fish from men must be the product of an accumulation of the minor variations that differentiate offspring from their parents. From this acceptance of gradualism it is only a tiny step to full-fledged Darwinism, since something (such as natural selection) must guide the process of descent with modification. The fact of evolution turns out to be gradual change through descent with modification from common ancestors, guided where necessary by natural selection, and this is Darwinism.

The Debate Lingers

The important debate is not between "evolutionists" and "creationists," but between Darwinists (scientific materialists) and persons who believe that purely naturalistic or materialistic processes may not be adequate to account for the origin and development of life. Once separated from its materialistic-mechanistic basis in Darwinism, "evolution" is too vague a concept to be either true or false. If I am told that the phyla of the Cambrian explosion evolved in some non-Darwinian sense from preexisting bacteria or algae, I do not know what the claim adds to the simple factual statement that the prokaryotes came first. It conveys no information about how the new forms came into existence, and the "evolution" in question could be something as metaphysical as the evolution of an idea in the mind of God.

Similarly, whether "creation" occurred over a greater or lesser period of time, or whether new forms were developed from older ones rather than from scratch, is not fundamental. The truly fundamental question is whether the natural world is the product of a preexisting intelligence, and whether we exist for a purpose that we did not invent ourselves. If Darwinists have not been overstating their case, they have disproved the theistic alternative, or at least made consideration of it superfluous.

A Comprehensive Worldview

Whatever may be its utility as a paradigm within the restrictive conventions of scientific materialism, Darwinism has continually been presented to the public as the factual basis for a comprehensive worldview that excludes theism as a possibility. A few representative quotations will suffice to make the point:

George Gaylord Simpson, in *The Meaning of Evolution*, states:

> Although many details remain to be worked out, it is already evident that all the objective phenomena of the history of life can be explained by purely naturalistic or, in a proper sense of the sometimes abused word, materialistic factors. They are readily explicable on the basis of differential reproduction in populations (the main factor in the modern conception of natural selection) and of the mainly random interplay of the known processes of heredity... Man is the result of a purposeless and natural process that did not have him in mind.[6]

Douglas Futuyma, in a citation written in a college textbook, states:

> By coupling undirected, purposeless variation to the blind, uncaring process of natural selection, Darwin made theological or spiritual explanations of the life processes superfluous. Together with Marx's materialistic theory of history and society and Freud's attribution of human behavior to influences over which we have little control, Darwin's theory of evolution was a crucial plank in the platform of mechanism and materialism—of much of science, in short—that has since been the stage of most Western thought.[7]

In his book *The Blind Watchmaker*, Richard Dawkins states:

> Darwin made it possible to be an intellectually fulfilled atheist.[8]

And, in *Evolution and the Foundation of Ethics*, William Provine has stated:

> The destructive assumptions of evolutionary biology extend far beyond the assumptions of organized religion to a much deeper and more pervasive belief, held by the vast majority of people, that non-mechanistic organizing designs or forces are somehow responsible for the visible order of the physical universe, biological organisms and human moral order.[9]

Has the Materialist-Mechanist Program Succeeded?

Whether the materialist-mechanist program has succeeded as the Darwinists have so vehemently claimed is a legitimate subject for intellectual exploration. Scientists rightly fight to protect their freedom from dogmas that others would impose upon them. They should also be willing to consider fairly the possibility that they have been seduced by a dogma that they found too attractive to resist.

With respect to the public schools, the providing of information about the evidence pertaining to Darwinism should be distinguished from efforts to indoctrinate students in "what scientists believe." Specifically, textbooks should be candid in acknowledging the origin of life problem, the fossil record problems, the limited results of selective breeding, and the inability to confirm experimentally the hypothesis that natural selection has creative power.

More importantly, the universities should be opened up to genuine intellectual inquiry into the fundamental assumptions of Darwinism and scientific materialism. The possibility that Darwinism is false and that no replacement theory is currently available should be put on the table for serious consideration.*

*"Darwinism on Trial," which has been called "The Campion Position Paper on Darwinism," is now published, with the same text but arranged in a numbered-paragraph format, on apologetics.org/articles. It is reprinted here by permission of the author, Dr. Phillip Johnson.

Phillip E. Johnson

Since 1990, the acknowledged leader of the Intelligent Design movement has been **Phillip E. Johnson,** a Harvard-educated law professor who taught for thirty years at the University of California at Berkeley. Johnson, first in his law class at the University of Chicago, was selected in the 1960s to clerk for Chief Justice Earl Warren of the United States Supreme Court. In his distinguished career as a law professor at Berkeley, his specialty was criminal law, criminal procedure, and legal ethics. He retired in 2000 to devote his energies to writing and speaking on the Design vs. Darwinism controversy and other cultural issues.

Besides his many guest columns in prestigious newspapers and contributed chapters in books on Intelligent Design, such as *Mere Creation* and *Signs of Intelligence*, Johnson has written six books. After he published his first critique, *Darwin on Trial*, he penned *Reason in the Balance*, *Defeating Darwinism by Opening Minds*, *Objections Sustained*, *The Wedge of Truth*, and *The Right Questions.* He is in demand as a leading lecturer on Design on university campuses in the U.S. and overseas.

For example, the Cambrian strata of rocks...are the oldest ones in which we find most of the major invertebrate groups. And we find many of them already in an advanced state of evolution, the very first time they appear. It is as though they were just planted there, without any evolutionary history.

—Richard Dawkins
From "Puncturing Punctualionism" in *The Blind Watchmaker*

Chapter 6

Of Canadian Oddballs and Chinese Monsters

By Thomas E. Woodward, Ph.D.

Let's play "Fossil Jeopardy"...

Recent discoveries about this ancient event were so significant that *Time* magazine published a cover story on them. As the writers struggled to adequately paint the picture of what happened, they dubbed it "biology's big bang." Dozens of newly unearthed creatures in China (alongside older Canadian fossils that also stunned researchers) were so striking and bizarre that *National Geographic* produced a feature article on the historic findings. Artists were commissioned to re-create several underwater scenes of ancient seas, teeming with these strange creatures.[1]

The Cambrian Explosion

By now, you may have guessed the mystery topic and blurted out the correct answer—which would be a "question" in *Jeopardy:* "What is the *Cambrian explosion?*" This striking name was given long ago by paleontologists to the sudden appearance of a rich variety of complex animal fossils in the lowest strata containing recognizable fossils. Most famous among these are the trilobites and the brachiopods, but the abundance of different types, even in Darwin's day, was striking. Virtually every modern phylum—every basic animal body plan—can be found in the Cambrian strata. What is most striking, these animals leap into the rocks abruptly, without a hint of transition from previous ancestors.

The Supreme Embarrassment of Darwin and His Disciples

In a sense, this is not out of the ordinary. To be blunt, the pattern of "sudden appearance," followed by "stasis" (stability of form, or nonevolution) is the characteristic pattern of fossils in every strata. Even in Darwin's day, this absence of transitions was a major stumbling block to many scientists, who were quite eager and persistent in pointing this out to Charles Darwin himself.[2] This obviously poses an embarrassing enigma for Darwinian evolution. If all creatures have developed from earlier, more primitive ancestors over eons of steady evolution, why do we never seem to find a trace of that development in the fossil record?

For several years I enjoyed an open and forthright correspondence with renowned Princeton biologist John Tyler Bonner, which entailed my asking many questions about the empirical problems of Darwinism.[3] When I asked about the general absence of transitional fossils, his first remark was, "Well, *it is quite interesting* that in the fossil record, new types of animals seem to appear in bursts." To answer my doubts, he then referred me to the book *The Meaning of Evolution* by Harvard evolutionist George Gaylord Simpson. Unfortunately, my reading Simpson gave no help; rather, it just confirmed my picture of Darwinism as a branch of science that was struggling with a gigantic quandary in this area.

This discontinuous pattern of fossils, now universally acknowledged by fossil experts, is surely not what Darwin would have predicted (or any of his disciples since then). In fact, it is the *dead opposite of Darwinian predictions.* To sum up: The fossil record is *the last place* to look if one wants to find good evidence that nature had accomplished the gradual resculpting of bodies into new forms—what scientists call "morphological evolution." What is morphological evolution?

To answer that, let me turn the clock back and relive a vivid memory from my high school days. It seems like yesterday that my lab partner and I prepared to dissect our first frog in freshman biology class at Canal Winchester High School in central Ohio. Dr. Judson Wynkoop, our beloved but tough high school biology teacher, drilled into us the importance of the "frog morphology" we were about to witness as we wielded our scalpels. In fact, all year long Doc Wynkoop had patiently ingrained into our thinking the two great "biological dimensions" of every creature on earth. One of these dimensions is the creature's "morphology," which is the physical structure or form of the body and of each particular organ, that makes it unique. This is the side of an animal or plant that can be most easily captured in a glimpse at the innards of a frog—or a fossil specimen. The other side of the creature, its physiology,* is much harder to glimpse from fossil studies. Thus, *evolution of*

* An animal's or plant's *physiology* relates to the *dynamic function* of the various parts (right down to cells and subcellular systems). That function can be described in terms of motion, growth, physical and chemical processes, and more. Examples of such processes that fit into the category of *physiology* would be digestion, endocrine function, or metabolism.

morphology is simply a "change in body structure." The profound embarrassment of macroevolution is right here—the fossil record seems to contain no hints at all of any pattern of change between basic morphologies or "body plans."

No one has done more to open up this mysterious pattern to public view (and to try to solve it with his theory of punctuated equilibrium) than the late American paleontologist Stephen Jay Gould.[4] To help us sense the wonder of the most massive of all the "sudden appearances"—to catch the sheer excitement of the Cambrian explosion—I shall let Gould himself describe this event for us. His book on this topic, *Wonderful Life*, is a gripping tale of the Canadian discoveries pertaining to this era, and I recommend it highly, but here I quote Gould in *Natural History:*

> Cambrian seas teemed with life preserved in an abundant fossil record. But when geologists...studied rocks of earlier, Precambrian times, they found nothing organic—not a trace of anything potentially ancestral to the diverse assemblage of trilobites, mollusks, brachiopods, and other creatures in Cambrian strata. This geologically abrupt transition from blankness to a rich fauna including representatives of almost every modern phylum has been called, in well-chosen metaphor, the "Cambrian explosion."[5]

Above, I emphasized that this abrupt debut of the richness of the Cambrian fauna had no connection with previous life. This raises a question: Was there any life in the rocks below, in the "basement" that underlies even the Cambrian strata? For over a hundred years (from Darwin's day until shortly after 1960), paleontologists thought that the pre-Cambrian rocks were completely devoid of fossils—like a blank sheet of paper. However, since the 1960s, scientists have learned that pre-Cambrian rock does contain a few scattered traces of life—mostly one-celled "microfossils," such as bacteria and blue-green algae. Also, a rare and unrelated set of creatures called the Ediacaran (or Vendian) fauna have been noted and catalogued. Many of them look vaguely like feathers rooted to the sea floor, and some bear a resemblance to a Frisbee with odd decorations.[6] Yet, since the Ediacaran animals are not considered ancestral to those of the Cambrian, and since bacteria clearly do not evolve abruptly into complex creatures like crabs and jellyfish, the pre-Cambrian discoveries have only heightened the mystery of the Cambrian big bang.

This fossil explosion has long been known and was even discussed in Darwin's day. In fact, this "sudden appearance" of the major types of life forms, in the lowest strata of rock, without any trace of origin from earlier progenitors, baffled Darwin greatly. To his credit, he even admitted in *The Origin of Species* (1859) that this "is probably the gravest and most obvious of all the many objections which may be urged against my views."[7]

The Burgess Shale: Discovery and Rediscovery

Fast forward to the eve of World War I. A celebrated American paleontologist, Charles Walcott, stumbled upon a Cambrian gold mine high in the Canadian Rockies in 1909. For seven years, digging summer after summer in a formation called the "Burgess Shale" in British Columbia, he unearthed some of the most beautifully preserved specimens ever seen from Cambrian rocks. These priceless fossils (which Walcott quickly classified into preexisting taxonomic categories) soon came into the possession of the Smithsonian Museum in Washington, D.C. Museum officials stashed away this precious cargo in safe storage bins, and then largely forgot about them for fifty years.

In the 1970s, a trio of British scientists based at Cambridge University began to study the Burgess Shale fossils, which had never been closely analyzed. They blew away the dust and began a thorough scrutiny of the morphology of each unusual creature in the collection. Soon, they were puzzling over unexpected body features and were astounded to find that several dozen of these animals were major discoveries—true "oddballs" hitherto unknown in the Cambrian explosion. In fact, many kinds of Burgess fossils fit no known category (or "taxon") of living thing. A striking example is the ugly little monster called "Opabinia." (See illustration below.) It had a streamlined body with many swimming-lobes (resembling tiny oars) along the sides of its body, a head with five protruding eyes, and sticking out in front, a proboscis-structure resembling a fire hose, with a grasping tip at its end. Such a creature was previously unknown among the already-rich Cambrian explosion.

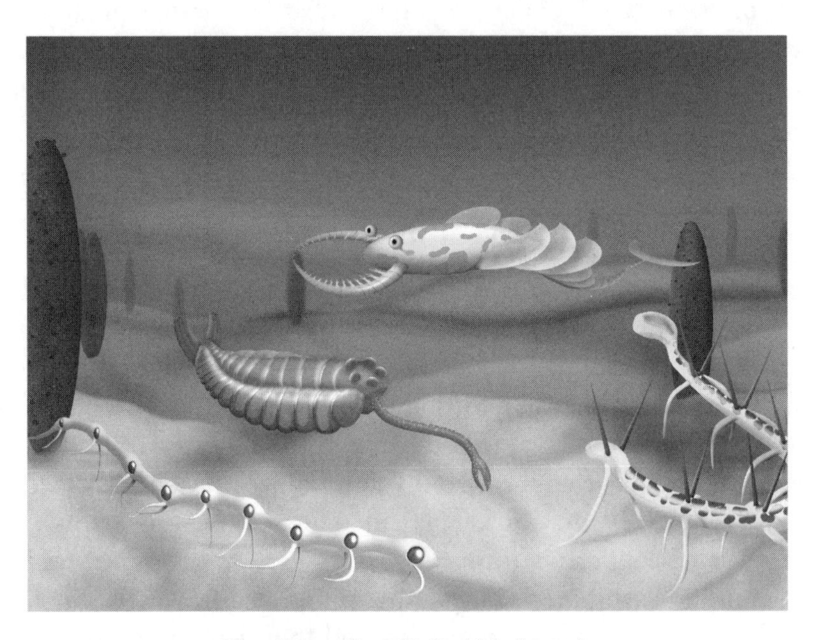

Microdictyon (far left), Opabinia (center), Anomalocaris (top center), Hallucigenia (right)

I could list and describe many other new Cambrian creatures that emerged from Walcott's Canadian gold mine, but here I must sum up. With the careful study and reclassification of the Burgess Shale oddballs in the 1970s (still ongoing today), the Cambrian explosion suddenly got significantly bigger. Because many of these creatures seem to fit no known phylum or class, the extreme "disparity"—the sheer difference in body plan from other Cambrian animals—is what struck many paleontologists. Gould pointed out in *Wonderful Life* that the popular textbook idea—of evolution producing a "cone of increasing diversity" in basic life forms over the ages—is now turned upside down. That is, the base of the cone (representing the time of greatest diversity) is at the bottom of the fossil record, in the Cambrian, not in the present. After that, no new basic body plans emerge; they just go extinct. The fossil experts at Cambridge University indeed had shocked the world of science, but as the first phase of their work wound down in 1985, they surely would not have dreamed that an even greater series of Cambrian surprises were about to come to light—across the world in Southern China!

Chinese Monsters Explode on the Stage of Paleontology

Dr. Jun-Yuan Chen is a tall, thin paleontologist who often sports his favorite blue cardigan sweater and flashes a friendly smile as he visits U.S. universities, lecturing in English about his amazing discoveries in China. When he is not digging or analyzing fossils, Dr. Chen teaches at China's prestigious Nanking Institute of Geology. In his early years at Nanking, even though he had experience in fossil studies, his main emphasis in teaching was in petroleum geology. However, in the mid-1980s, one of his former students arrived one day for a visit to show his professor a beautifully preserved Cambrian fossil he had dug up in South China. The fossil immediately caught his attention, and Dr. Chen asked the student where he had found it. "In Yunnan Province," said the student, "near the village of Chengjiang" (pronounced "cheng-jang"). This village is one of many that cluster in a sleepy, rolling agricultural countryside in Yunnan, peppered with hills and lakes.

Before long, the student led Dr. Chen to the exact spot where the fossil was found. Chen and his colleagues immediately began digging in this area and were stunned with the rich finds they were unearthing. By the early 1990s, thousands of precious Cambrian specimens had been pulled from the fine yellow shale, which had acted to preserve even the delicate *soft-bodied Cambrian animals* in exquisite, almost incredible detail. In some cases, the exact shape and contents of interior organs were revealed. What the student had led Dr. Chen into, to put it in historical terms, was the greatest Cambrian fossil bonanza of all time—eclipsing even the Burgess Shale fossils.

In the past twenty years as the digs at Chengjiang have spread to more and more sites in this rural district of southern China not far from the Vietnam border, the many thousands of Cambrian fossils unearthed so far have begun (again) to shake up the world of paleontology. Just as the Cambridge scientists were astonished to find so many new types pulled from the Burgess Shale, so also in China, a fountain of fresh discoveries—some of the weirdest new species yet—has poured out of the Chengjiang fossil beds as well. Let me introduce you to a few of them.

Hallucigenia is a good candidate for the weirdest Cambrian fossil to come to light in recent decades. It was a small worm-like creature with a unique set of appendages. (See page 100.) When the Cambridge scientists who perused the Canadian fossils first studied this animal, they noted on one side seven pairs of needlelike spines and, on the other side, seven tentacles. They were so astonished by this never-before-seen morphology, the word *hallucination* came to mind. As discussion began as to how to name the creature, the whimsical name "Hallucigenia" was suggested and quickly chosen.

Researchers puzzled about the seven pairs of spines, speculating that the worm walked on them, like stilts, in a strange stiff-legged mobility. Only after the same creature turned up in the Chengjiang fossils did the mystery of the "stilt walk" begin to be solved. Dr. Chen's team, excited to find Hallucigenia among their specimens, too, noted something new in their scrutiny of one such fossils. (See page 100). There appeared quite clearly in this specimen, that opposite each pair of needle spines, there was not just one tentacle, but a pair of tentacles! Thus, overnight, the Chinese evidence showed that the previous artist's conceptions of Hallucigenia were upside down. That is, in all likelihood, the animal walked on the seven pairs of tentacles and sported the seven pairs of needles on its back as protection. (Of course, Hallucigenia is extinct, which may strike biologists as a tragedy, but beach bathers around the world consider it a blessing.)

Microdictyon (pictured on page 100) is one of the most famous Chinese surprises. Its unique feature is a row of "eyes" that it sported on each side of its body, one alongside each leg where it joined with the worm-like body. Fossil experts, working with other biologists, can only speculate what these structures were. They are not at all sure what purpose they served, but because they look like eyes, that's what they're called!

The award for "most bizarre and fearsome" of all the new Chinese Cambrian species should go to something called *Anomalocaris.* As you can see from the illustration on page 100, this nightmarish beast had a large main body shaped somewhat like a flying saucer (or a spaceship from *Star Trek*), with a large tapered tail lined with swimming-lobes. Protruding from the top of its head near the front was a pair of large eyes on stalks, and reaching out in front was a pair of grabbing-and-feeding arms that vaguely resemble freshly caught jumbo shrimp. Originally, the arms were considered separate species, since they were found separately, broken off from the main body. Then, both the Canadian and Chinese fossils revealed what the complete body looked like, and the puzzle was solved.

When Dr. Chen came to Florida recently, it was my privilege not only to travel with him as he visited major colleges and universities (acting as chauffeur, valet, and understudy), but also to see firsthand the spectacular Chengjiang fossils he brought with him.

I recall especially the tiny yellowish fossil, about four inches in length, in which one could see in dramatic detail the entire body of an Anomalocaris. I asked if this was a "full-grown" specimen. Dr. Chen laughed, "Oh no, that is a baby. We have found some that were five or six feet in length!" Once again, I was secretly happy that this sea monster has gone extinct.

Summing Up...So What?

Both the Canadian and Chinese fossil discoveries constitute one of the great modern true stories of paleontology. And that story is still unfolding as more and more unique, never-before-seen creatures are being dug from Chengjiang each year. But for the issue of Darwinism and design, it is a story of a grand but aging scientific paradigm that has ruled biology with a firm hand for one hundred fifty years, finally meeting its match in the fossil evidence. The Cambrian explosion, ever a headache and mystery for true Darwinians, is much worse now than it was thirty years ago. These shocking fossils—so rude, so inconvenient, so recalcitrant—have utterly defied the conventional notions of naturalistic evolutionism.

Ever since Darwin, a whole list of explanations, excuses, and ad hoc hypotheses has been deployed to explain the puzzle of missing transitions. However, each of these fails utterly to offer any escape route from the newly enlarged Cambrian explosion. For example, one often hears, "Perhaps the ancestors to the Cambrian animals were soft-bodied, and thus they were not preserved." In Chengjiang, we find the soft-bodied animals beautifully preserved alongside those with hard parts, and recent findings show that soft-bodied animals are also preserved in the pre-Cambrian.

Alternatively, a follower of Gould's idea of punctuated equilibrium may say, "Perhaps the ancestors of the Cambrian phyla were tiny animals, hard to recognize as ancestors, and they moved to the periphery of their population, were isolated, then evolved with incredible rapidity due to macromutations." Indeed, Gould offered an explanation somewhat like this in his book on the Burgess Shale, calling it the "fast-transition" theory. Yet, as Phillip Johnson has pointed out cogently, Gould's words "fast-transition" lack any solid empirical content. Rather, they are like a scientific label, stuck on the top of a mystery and thus giving the unwary onlooker the superficial appearance of knowledge.

Clearly, now that we may wind up (according to Chen's estimate) with as many as ninety phyla in the Cambrian explosion, Darwinism has reached a severe crisis in terms of *genuinely plausible explanation.* Obviously, ninety complex phyla—with morphology as strange and varied as Opabinia, Hallucigenia, and Anomalocaris—did not suddenly evolve from those bacteria or algae without leaving a trace of their drastically changing transitional intermediates!

On several occasions as I accompanied Dr. Chen on his tour of universities in Florida, I heard him conclude his lecture by saying, "Darwinism cannot explain the Cambrian explosion any more. We need a new theory." In every case, he was greeted by embarrassed silence. In one of the classes I teach at Trinity College, I interviewed Dr. Chen in front of my class and asked him if he meant by this statement that we need a new paradigm—a whole new perspective or system of thought. "Yes, in my opinion, we do," he replied, "and it will be all the more difficult, because the Darwinian thought is in a box."

I am not implying that Dr. Chen has completely given up on the idea of common ancestry. It is reported that his team is exploring a variety of ideas—including the notion of intelligent design as outlined by Dr. Behe and his colleagues in the Design Movement. What Dr. Chen *has given up on*, without a doubt, is the "engine of Darwinism"—the idea that random mutations in DNA, filtered by natural selection, can slowly change and sculpt new bodies, new species, and even new phyla. The Cambrian explosion—now expanded by *Time* magazine into the "biology's big bang"—is the deathblow to the Darwinian scenario, in Chen's view.

Dr. Chen left Florida and returned to California, where he was based in San Francisco for many months. He continued to lecture in universities in that state, sharing the same grim conclusion about Darwinism. As Dr. Chen finished his United States' tour and prepared to leave for China, he turned to the biologist who was taking him to the airport and said, "Why does no one comment when I tell audiences that Darwinism won't explain the Cambrian data?" The scientist quickly explained that this area was extremely sensitive to American scientists, and that any strong hint that Darwinian theory was in trouble was virtually taboo in the universities.

Dr. Chen laughed, "This is very interesting. In my country, you can criticize Darwin, but not the government. In your country, you can criticize the government, but not Darwin!"

Perhaps the day is not far away when the intellectual freedom of speech in the United States will catch up with that of our Chinese friends. The new Cambrian monsters are surely pushing us in that direction.

Thomas E. Woodward

A graduate of Princeton University and Dallas Theological Seminary, **Thomas E. Woodward** received his Ph.D. in 2001 at the University of South Florida in Communication, specializing in the Rhetoric of Science. His dissertation, "Aroused from Dogmatic Slumber," was a rhetorical history of the Intelligent Design Movement.

He is the chair of Theology and Biblical Studies at Trinity College of Florida and is founder and director of the C. S. Lewis Society, a nonprofit organization that hosts on university campuses lectures and seminars pertaining to ethics and apologetics. This included the organization of seminars on the Yale and Princeton campuses on the evidence and reasons that undergird the Christian worldview. He is the creator and director of the Apologetics.org website, which serves thousands of visitors each month from around the globe.

Dr. Woodward has published numerous articles on the creation-evolution controversy and is a frequent speaker on the topic at universities in the U.S. and overseas, including the countries of Greece, Yugoslavia, Hungary, the Dominican Republic, Cuba, Uruguay, Mexico, the Netherlands, and Romania. He is also the producer of the Princeton Series of videotapes, including *The Princeton Chronicles*, *Opening Darwin's Black Box*, *A Chemist's Story*, and *Is There a Rational Morality?* He has also ventured into the production of videotapes in the arena of global missions, producing *The Duna Story* and *Light of the Nations.*

Before coming to Trinity College in 1988, Dr. Woodward was a missionary in the Dominican Republic, working with college students and professors in Santo Domingo. He now lives with his wife, Normandy, a schoolteacher, in Tampa Bay.

Part Two

Irreducible Complexity and the Detection of Design

If it could be demonstrated that any complex organ existed which could not possibly have been formed by numerous, successive, slight modifications, my theory would absolutely break down.

—Charles Darwin, in *The Origin of Species*

To Darwin, the cell was a "black box"—its inner workings were utterly mysterious to him. Now, the black box has been opened up and we know how it works. Applying Darwin's test to the ultra-complex world of molecular machinery and cellular systems that have been discovered over the past 40 years, we can say that Darwin's theory has "absolutely broken down."

—Michael Behe, biochemist and author of *Darwin's Black Box*

SEVEN

Chapter 7

Meeting Darwin's Wager

BY THOMAS E. WOODWARD, PH.D.

During the fall of 1996, a series of cultural earthquakes shook the secular world with the publication of a revolutionary new book, Michael Behe's *Darwin's Black Box: The Biochemical Challenge to Evolution.* The book reviewer in the *New York Times* praised Behe's deft analogies and delightfully whimsical style and took sober note of the book's radical challenge to Darwinism. Newspapers and magazines from Vancouver to London, including *Newsweek*, *The Wall Street Journal*, and several of the world's leading scientific journals, reported strange tremors in the world of evolutionary biology. *The*

Chronicle of Higher Education, a weekly newspaper read primarily by university professors and administrators, did a feature story on the author two months after his book appeared. The eye-catching headline read, "A Biochemist Urges Darwinists to Acknowledge the Role Played by an 'Intelligent Designer.'"

Now reporters are making their pilgrimage to Bethlehem, Pennsylvania, to interview the author at the center of these quakes—Lehigh University biochemist Michael J. Behe (pronounced "bee-hee").

Behe, who typically sports a lumberjack shirt, jeans, and black Adidas sneakers, toils long hours with his students in the biochemistry lab, doing research on DNA and the structure of proteins. He is short, balding, and has thick, dark-rimmed glasses; he looks as much like a hardware-store clerk as a scientific renegade.

Seated at a lab table, surrounded by bottles filled with clear, smelly fluids designed to rearrange DNA sequences, he explains that advances in his own field—where scientists have been furiously unraveling the mysteries of exactly how cells work—have yielded a startling finding. Molecular machinery and complex systems in the cell are dependent upon far too many interconnected parts to have been built up gradually, step by tiny step, over time.

With his book already in its thirtieth printing, Behe finds his calendar filling up with speaking engagements. In a recent trip to the University of South Florida in Tampa, he spoke to biologists, students, and schoolteachers who had braved rains from an approaching hurricane to hear him.

In his talk, Behe quickly reviewed the modern theory of evolution and then flashed onto a screen his favorite quote by Darwin from *The Origin of Species*, acknowledging the kind of evidence that would be necessary to refute the Darwinian theory of evolution.[1]

Behe took up the challenge of Darwin's test and asked, "What type of biological system could not be formed by numerous, successive, slight modifications? Well, for starters, a system that has a quality that I call irreducible complexity."

Encouraging the nonscientists in the audience to stay tuned, Behe explained briefly what he meant by the phrase. "When I say that something is irreducibly complex, I simply mean it is a system composed of several well-matched, interacting parts that contribute to the basic function, wherein the removal of any one of the parts causes the system to effectively cease functioning."

With his characteristically impish grin breaking through a full beard, Behe flashed on the screen a diagram of the humble mousetrap, his trademark illustration of "irreducible complexity." After pointing out the five parts necessary for mousetrap function, he added, "You need all the parts to catch a mouse. You can't catch a few mice with a platform, then add the spring and catch a few more, and then add the hammer and improve its function. All the parts must be there to have any function at all. The mousetrap is irreducibly complex."

Behe was suddenly a tour guide, piloting his listeners on a theme park ride through the cell and pointing out systems that exhibited this eerie mousetrap kind of complexity. Using photos and diagrams, he walked through the chemical chain reaction that gives rise to vision and detailed the elegant but complex structure of the whiplike cilium with which many kinds of cells are equipped. *Far Side* and *Calvin and Hobbes* cartoons punctuated the lecture, and even an outlandish Rube Goldberg contraption—the "Mosquito Bite Scratcher"—was displayed as an analogy to the complicated mechanism by which blood clots form.

"The cell is no longer a mysterious black box as it was for Darwin," Behe continued. "We now know precisely how it works at the molecular level. And the cell is chock full of systems like these that are irreducibly complex."

Finally, he showed a *New Yorker* cartoon with a professor being confronted in his office by his department chairman and by a hit man who is screwing a silencer onto his gun. The caption reads, "Surely, professor, you knew when you took this position, it is publish or perish!"

His listeners relished the humor, but the mood in the room turned serious as Behe made his point:

> As you search the professional literature of the last several decades, looking for articles that have been published even attempting to explain the possible Darwinian step-by-step origin of any of these systems, you will encounter a thundering silence. Absolutely no one—not one scientist—has published any detailed proposal or explanation of the possible evolution of any such complex biochemical system. And when a science does not publish, it ought to perish.

In short, Behe said, modern evolutionary theory, applying Darwin's own test, flunks spectacularly at the molecular level. Rather, everywhere we look inside the cell, evidence is staring scientists in the face that suggests the systems were directly designed by an intelligent agent.

Michael Behe is the father of six children, three boys and three girls ranging in age from two through eleven, with a seventh on the way. No wall of separation stands between his fathering and his writing about biochemistry. He weaves into many of his chapters homey images drawn from the Behes' family room. For example, the joyous task of assembling his son's tricycle on Christmas Day illustrates the importance of detailed instructions in living systems. Putting together snap-lock beads and Tinkertoys with his kids on the family-room rug provides pictures of how organic molecules are built. His youngest daughter's dolly wagon is pressed into service to help explain how antibodies latch onto the body's invaders. Behe, the master teacher, can hardly make a point without bringing in something familiar and concrete, such as tuna cans, an elephant, chocolate cake, and even "road kill."

Behe's wife, Celeste, has a teaching career as well; she homeschools four of the Behe children. When Mike Behe comes home from the biology lab, he enjoys playing Frisbee with his children and reading to them. In fact, the Behe home is like a library—children's books are scattered throughout the house and are stacked on sixteen shelves in a room set aside especially for reading. Since the Behes decided seven years ago not to keep a TV in their home, the Behe children have found time for reading good books, learning karate and piano, and singing in the church choir.

Mike Behe's efforts in fathering his children are balanced by his new fathering task in his own scientific field. One could describe *Darwin's Black Box* as a "birthing book"—it is Behe's proposal to give birth to a new perspective in biology that stops ignoring the pervasive presence of "design." He is not alone in the task. Behe has worked closely with an interdisciplinary team of scientific colleagues scattered in colleges and universities from Seattle to Princeton, New Jersey.

The acknowledged leader of the "Design Movement" is Phillip Johnson, a law professor at the University of California at Berkeley, whose *Darwin on Trial* (revised 1993) has stirred vigorous interaction with the world's most prestigious evolutionists, including Stephen Jay Gould of Harvard University and Niles Eldredge of the American Museum of Natural History.

According to Johnson, Behe's book has inaugurated a new phase of the critique of Darwinism. Behe not only devastates the case for Darwinism at the molecular level, but he is also leading the way in fashioning a new frame of reference on origins.

The goal of the Design Movement is to liberate science from its shackles of naturalistic philosophy so that scientists who probe the origin of nature's wonders will have the freedom to consider all the possible explanations—including design by an intelligent agent. An international conference was held in November 1996 at Biola University in Los Angeles, drawing together 180 college professors and other researchers to consider a revolutionary proposal for new scientific and mathematical principles that can help determine how something in nature arose.[2]

The basic idea is to ask, "Which of three possible explanations fits best in explaining a given phenomenon, X? Can X be explained by lawlike actions of nature, or could X be the result of random events, or, failing these possibilities, is X the result of action by an intelligent agent?" This three-way test (dubbed "the Explanatory Filter") became the centerpiece of the conference as Behe and his colleagues reviewed new evidence that points to design. Some observers say that the Design Movement may be embarking upon the first stage of a transitional process in science, which philosophers call a "paradigm shift."

In his book, Behe argues that the time has come for biological science to face the logical implications of what it has been finding in biochemistry and to get down to an important new task: identifying which contraptions in the cell clearly bear the marks of intelligent design, and which ones could have developed from earlier systems.

The timing for such a revolutionary turn seems to be right, as is suggested by the furor provoked by the June 1996 issue of *Commentary* magazine. The lead article of that issue was "The Deniable Darwin," a sophisticated skewering of Darwinism by Princeton-trained philosopher and logician David Berlinski. Under the title appeared the provocative line, "The fossil record is incomplete, the reasoning flawed; is the theory of evolution fit to survive?"

Commentary published in its September issue an astonishing thirty-three-page section devoted to the wave of responses to Berlinski's article. Angry letters had poured in from the world's leading Darwinists, but other scholars praised the author for his rigorous analysis and the editors for their intellectual courage in publishing the piece. The author took thirteen pages to respond, point by point, to each letter.

Berlinski, author of the recent award-winning book *A Tour of the Calculus*, says that skepticism regarding Darwinian orthodoxy has now exploded out of its evangelical Protestant ghetto and that revolution is in the air. He points to Behe's work as a turning point in this process: "*Darwin's Black Box* is simply an extraordinary piece of work that will come to be regarded as one of the most important books ever written about Darwinian theory. No one in the evolutionary camp can propose to defend Darwin without meeting the challenges that Behe has set out in his book—it's really quite compelling."[3]

Instead of ignoring Behe, as many tried to do to Phillip Johnson, both the media and the scientific establishment are paying close attention to the feisty biochemist at Lehigh. The treatment accorded Behe in the *New York Times*, "the paper of record," is one sign of this cultural shift. The first significant notice came on August 4, 1996, when *Darwin's Black Box* was honored by a review in the *New York Times* book reviews. Evolutionist James Shreeve expressed appreciation for Behe's knack of explaining natural wonders. In the end, Shreeve did not agree with Behe's intelligent design proposal, saying we should not jump the gun and say, "God did it," but rather leave some mysteries for our grandchildren to work on. But the review conveyed Behe's thesis clearly:

> He argues that the origin of intracellular processes underlying the foundation of life cannot be explained by natural selection or by any other mechanism based purely on chance. When examined with the powerful tools of modern biology, but not with its modern prejudices, life on a biochemical level can be a product...only of intelligent design. Coming from a practicing scientist...this proposition is close to heretical.

Even more noteworthy was the appearance of Behe's own article, "Darwin Under the Microscope," in the op/ed pages of the *New York Times* (Oct. 29, 1996). The steps leading to this began in mid-September when an editor of the *Times* startled Behe by asking if he would consider submitting an article explaining the main theses of his book.

In reply, Behe immediately wrote an op/ed piece, which lingered for a month on an editorial desk. Then, on October 25, front-page headlines around the world reported Pope John Paul II's puzzling (and widely misunderstood) statement on evolution as being "more than a hypothesis" based on "fresh knowledge" that scientists should be free to investigate, keeping in mind that the soul is a direct creation of God.[4]

Since Behe is a Roman Catholic scientist who teaches in the biology department at a major university, both the *Times* and Behe sensed a tie-in. Within a day, his piece had been rewritten to connect it with the pope's statement.[5]

In this article, Behe explained that, for him, the pope's statement is nothing new. As a Catholic, Behe was taught that evolution could be viewed as God's way of creating.

What forced Behe to change his mind about the truth of Darwinism and to propose intelligent design was not religion, but scientific discoveries in his own field. The pope spoke of "several theories" of evolution, Behe noted, explaining that the only valid theory of evolution that he saw emerging from the biological evidence took note of the unmistakable signs of "intelligent design."

Inevitably, many scientists charge Behe with "thinly disguised creationism." This strategy is employed by University of Chicago biologist Jerry Coyne, whose review of Behe was published in September in the prestigious British journal *Nature*. While Coyne admits, "There is no doubt that the pathways described by Behe are dauntingly complex and their evolution will be hard to unravel," he claims that Behe has offered no solution: "Behe's 'scientific' alternative to evolution [is] a confusing and untestable farrago of contradictory ideas." Twice in the review Coyne's rhetoric links Behe to the San Diego "scientific creationists" whom professional evolutionists tend to dismiss. Coyne describes Behe's work as a "new and more sophisticated" version of literal-Genesis creationism.[6]

In fact, Behe has explained clearly his differences with the young-earth creationists. For example, he is willing to accept "as a working hypothesis" Darwin's concept of common ancestry. He even declares, "I am not a creationist," defining the word narrowly as including a belief in recent six-day Creation as derived from a literal reading of Genesis.

Behe believes that God happens to be the intelligent Designer to which his biochemistry findings are pointing, but he stresses that science itself may not have the ability to ferret out the identity of the designer any more than astronomers can determine from their measurements the one who caused the expanding universe to spring into being out of nothing. Behe sees science and religion as two lines of investigation that connect or overlap in the area of origins, but neither of these human endeavors can claim to usurp the function of the other.

Thus, religion may help create the conceptual space needed for Behe's thinking to change, but he traces his doubts about Darwin to a series of intellectual shocks or "rude scientific awakenings" he received while working in the arena of biological origins over the past decade. His thinking took unexpected turns through interactions with colleagues in biochemistry, whose contagious skepticism of Darwin stirred him to carry out his own investigations, which in turn led to his emergence as a leading figure in the design movement.

Michael Behe grew up in Harrisburg, Pennsylvania, as one of eight children in a middle-class family. His father, taking advantage of the GI Bill, was the first of his family to go to college, and he became the manager of a branch of Household Finance Corporation.

Even as a child, Mike Behe says, he was a "science enthusiast." He excelled as a high school student, graduating fifth in a class of two hundred and being elected senior class president. Recalling his Catholic high school science classes, he says, "I was taught that God made the laws of the universe and that some of those laws led to evolutionary processes. Therefore, God is no less a creator just because He uses the laws He set in motion."

To Behe, evolution was never a point of contention until he reached Philadelphia's Drexel University in the early 1970s. He vividly recalls an odd conversation with a fellow student who used evolution as a "tool to fight religion." Behe argued vigorously with this campus skeptic for the theistic position on evolution, but when the dust of battle had settled, neither had converted the other. In

1974, Behe graduated from Drexel with a degree in chemistry and an education in the uses of Darwinism for propaganda in the hands of atheists.

For his doctoral studies, Behe moved across town to the University of Pennsylvania. There he plugged away for four years and, after completing his Ph.D. in biochemistry in 1978, attained an appointment to the National Institutes of Health in Bethesda, Maryland.

One of his colleagues in the genetics laboratory at the National Institutes of Health was a fellow Catholic biochemist, Jo Ann Nichols. Rarely did their work touch on evolution, but Behe recalls one day when the issue did arise as a matter of joint speculation between them during a break. The question was this: "If the first life did arise by random naturalistic processes from a chemical soup, as all textbooks are saying, what exactly are the minimum systems that are required for life?" Together they ticked off a mental list of the minimum requirements: a functioning membrane, a system to build the DNA units, a system to control the copying of DNA, a system for energy processing. Suddenly, they broke off their speculation, looked at each other, and smiled, jointly muttering, "Naaah—too many systems; it couldn't have happened by chance."

In 1982, Behe was hired by Queens College in New York City to teach biochemistry. He looks back on the three years at Queens as a high point of his life, primarily for what happened outside the lab and the classroom. It was while living in Queens that he met his wife, Celeste, a bright, attractive young woman with jet-black hair who had grown up in an Italian Catholic family. After a three-month courtship, Michael proposed, and they were married the following summer.

Three years later, not wishing to raise his family in an urban setting, he began to look elsewhere. When a position opened at Lehigh, an hour north of Philadelphia in Bethlehem, he applied and was brought onto the faculty in 1985, receiving tenure two years later.

It was shortly after his tenure was granted that he experienced the first major intellectual shock concerning evolution when he ordered the controversial book *Evolution: A Theory in Crisis* by agnostic geneticist Michael Denton.

As Behe opened the book, he found himself pulled in by Denton's radical scientific critique, which, while agreeing that "microevolution" is an established fact that no one denies, challenges the really significant claim of Darwinism—that it has explained "macroevolution." Denton, who now researches human genetics at Otago University in New Zealand and is not himself a creationist, defines macroevolution as the emergence of wholly new organs or organisms by purely naturalistic processes that work in small increments. Having evaluated the evidence for macroevolution and found it wanting, Denton concludes: "The Darwinian theory of evolution is no more nor less than the great cosmogenic myth of the Twentieth Century."[7]

Reading Denton's book was a "scientific wake-up call" for Behe. The intellectual effect, he says, was roughly analogous to an electroshock treatment, convincing him that the creative power of natural selection was mostly *bluff*—a largely unwarranted inference that was not well supported by the available evidence. Soon he was rethinking everything he had been taught about evolution, especially in his own specialty of biochemical systems.

In 1989, Lehigh's dean of the College of Arts and Sciences sent out a memo asking professors to develop proposals for new freshman seminar courses that explore exciting topics at the frontiers of knowledge in order to help students develop critical thinking skills. Behe saw this as a golden opportunity and submitted an outline for a course called "Popular Arguments on Evolution." His course employed three primary texts: Besides Denton's critique, he required students to read Thomas Kuhn's classic, *The Structure of Scientific Revolutions*, and a recent bestseller, *The Blind Watchmaker*, a defense of Darwinism by Oxford biologist Richard Dawkins.

Behe's proposal was accepted, and he has taught the course

almost every year since 1989. Throughout the course he lines up, side by side, evidence and arguments both for and against the conventional theory of evolution. His goal is to teach it so that the students will not necessarily know his personal position on macroevolution.

Nevertheless, he is pleased by his students' responses to his survey of the evidence. "I find it gratifying," says Behe, "how many students come to me at the end of the course each year and say, 'Prof, thanks for a great course; before, I had no idea that there was any scientific case against Darwinism.'"

During the same time, at the University of California at Berkeley, Phillip Johnson was polishing his own critique of Darwinism. Originally presented to a colloquium of his fellow professors, *Darwin on Trial* finally made it into book form in 1991. Johnson's four-year study of the scientific basis of evolution had also been triggered by reading Denton. Now his own book moved beyond Denton, not only reviewing the weak state of the scientific evidence for macroevolution but also pinpointing the key roles that philosophical assumptions were playing in the presentation and defense of Darwinism.

In July 1991, Mike Behe opened a copy of the journal *Science*, published by the American Association for the Advancement of Science. In its news column "Briefings," the issue reported on *Darwin on Trial* in a way that weather forecasters post hurricane warnings.[8] In effect, the news column dismissed Johnson as a know-nothing lawyer who misunderstood "how science works" and warned the readers that his book posed a danger to good scientific thinking. Eugenie Scott, director of the Anti-Creationist National Center for Science Education, fretted about Johnson's potential influence: "I hope scientists find out about this. They really need to know [the book] is out there and is confusing the public."

Behe had just begun reading *Darwin on Trial* and was furious with what he calls a "profoundly anti-intellectual" attitude toward

Johnson's work. Behe sent in a wittily barbed reply to *Science*, which they printed in a subsequent issue.[9] His letter has become a tiny classic in the literature of skeptics of Darwinism, and it immediately brought Behe to the attention of the design movement.

In late 1991, the Foundation for Thought and Ethics (FTE), a Dallas think tank, began organizing a symposium around Johnson's new book, to be held in March 1992. The idea was to invite five Darwinists and five proponents of intelligent design to debate the central thesis of *Darwin on Trial*—namely, that Darwinism is fundamentally grounded in philosophical preference, not scientific inference. Behe accepted FTE's invitation to join the Intelligent Design side, yet he admits that he entered the conference room at Southern Methodist University in Dallas with "some trepidation." Says Behe, "I just didn't know what to expect, nor did the Darwinists. Nothing like this had ever been attempted at a high academic level."

Right away the apprehensions of Behe and the others melted, and three days later all eleven participants left Dallas saying that the symposium was one of the best they had ever attended in their academic careers. "There were no conversions on either side," recalls Behe, "but a genuine spirit of camaraderie and mutual acceptance grew among us. It was one of the highlights of my life."

The proceedings, published under the title *Darwinism: Science or Philosophy?*, were hailed as a scientific milestone in the renowned journal *Quarterly Review of Biology*.[10] The volume contains a debate between Phillip Johnson and Darwinist philosopher of science Michael Ruse, along with the ten papers that were presented at the conference. Each paper is followed by a published reply by one of the participants on the other side of the issue.

Many observers described Behe's paper on the isolated nature of protein families as a "scientific bombshell." Using statistical and biochemical analysis, Behe proposed that the informational structure of proteins points to an intelligent designer, just as a book's letters must be formed in correct order by an author to produce coherent text. Yet what many recall as Behe's high point was his scintillating reply to an impressive Darwinist paper dealing with the immune system. The polite but scientifically high-octane contributions of Behe were a highlight of the symposium.

A year later the Johnson-Behe cadre of scholars met at Pajaro Dunes on the California coast. Here, Behe presented for the first time the seed thoughts that had been brewing in his mind for a year—the idea of "irreducibly complex" molecular machinery.

Once Behe had signed a contract with the Free Press, he proceeded to tap out the book's text on his computer. Behe stumbled onto a major surprise during his final stages of research as he began surveying college biology texts and technical journals—he previously had no idea how many Darwinian explanations for complex systems would turn up in the literature. He suspected that such proposed explanations would be few and far between, but what he found was far more eloquent—a total, systematic absence of any attempts. His excitement grew month by month as his search confirmed the universal silence on the topic.

In late July 1996, Mike Behe sat down in his office, flicked on his computer, and began paging through his e-mail messages. It had been an exhilarating month—his book was finally rolling off the press. He was excited about how well his half-day press briefing had gone in Washington, D.C., in front of dozens of intellectuals and media persons. While vacationing at the Maryland shore with his family, he had received an overnight package from Free Press that contained a copy of his literary first-born. Then, a few days later, word had come that a review would appear in the *New York Times*. That news brought excitement mingled with dread—he felt like celebrating, but he wondered if he should brace for an attack.

As Behe scanned the e-mail list, he spotted a message from Phillip Johnson. As he clicked open Johnson's message and scrolled through it, he smiled at his pep talk: "Don't worry, Mike. Even if the *Times* bashes you in their review, a cultural earthquake will take place [in the United States] on August 4 when they publish it."

A few days later, Behe received an early copy of the review and typed an e-mail report that popped onto computer screens of several dozen colleagues in the Design Movement: "Good news—I just got the *New York Times* review. Not bad. Not bad at all. On a scale of one to ten (ten being ecstatic praise, one being a total bashing), it's an eight." Behe could already feel the distant tremors.

As Behe lectures, one of the first questions asked is, "What do Darwinians say about your book?" He ticks off three or four recurring responses. A few simply label him a "creationist" and dismiss his arguments without a careful hearing, but that is not the typical response. Almost all reviewers have admitted that Behe has the facts right. Biochemist James Shapiro said that *Darwin's Black Box* had actually understated the complexity of the cell's systems, while James Shreeve conceded that "Behe may be right that given our current state of knowledge, good old Darwinian gradualist evolution cannot explain the origin of…cellular transport."

Nevertheless, Shreeve and others say the professor from Lehigh simply has given up too soon. Many add that science simply cannot entertain such unscientific notions as "intelligent design." Behe considers this objection a transparent attempt, based on philosophical bias, to set limits on science.

Some critics have sought refuge in the new mathematically based ideas of Stuart Kaufmann, a professor at the University of Pennsylvania who uses computer models to simulate what he calls "spontaneous ordering of life." Behe critiques Kaufmann's ideas in his book, pointing out that a recent article in *Scientific American* described Kaufmann's work as a "fact-free science." Behe emphasizes that Kaufmann's models never refer to real chemical or biological data and have produced no laboratory experiments. Thus, he concludes, Kaufmann's ideas offer no hope as an escape route for the Darwinians.

After reactions to *Darwin's Black Box* had poured in from professional biologists, Phillip Johnson noted, "All the criticism of Behe's book so far doesn't challenge the truth of what he says. It just reflects how unhappy it makes the Darwinists to see the scientific evidence and their materialist philosophy going in opposite directions."

This unhappiness was evident at the recent University of South Florida lecture. The professor who teaches the university's undergraduate course on evolution objected, "You're giving up too soon. Biochemistry is in its infancy. These systems were discovered just twenty or thirty years ago. Within the next few years, we may begin to figure out how all these systems evolved."

Behe replied, "Actually, many of these systems have been fully understood for forty years or more, and not one explanation has been published offering a plausible scenario by which they could have evolved. Any science that claims to have explained something, when in fact they have published no explanation at all, should be brought to account."

Michael Behe really wants to be nothing more than a biological accountant, initiating a long-overdue audit of the Darwinian books. The world is watching the results.*

---

*"Meeting Darwin's Wager," which was first published in *Christianity Today*, April 1997, is reprinted here by permission of the author, Dr. Thomas Woodward.

The result of these cumulative efforts to investigate the cell—to investigate life at the molecular level—is a loud clear, piercing cry of "design!" The result is so unambiguous and so significant that it must be ranked as one of the greatest achievements in the history of science.

—Michael Behe
From "Science, Philosophy, Religion" in *Darwin's Black Box*

Chapter 8

The Modern Intelligent Design Hypothesis: Breaking Rules

—By Michael J. Behe, Ph.D.
Department of Biological Sciences
Lehigh University, Bethlehem, Pennsylvania

In this chapter I will argue that some biological systems at the molecular level appear to be the result of deliberate intelligent design (ID). In doing so I am well aware that arguments for design in biology have been made before, most notably by William Paley in the nineteenth century. So I think it is important right at the beginning to clearly distinguish modern arguments for intelligent design from earlier versions. The most important difference is that my argument is limited to design itself; I strongly emphasize that it is not an argument for the existence of a benevolent God, as Paley's was. I hasten to add that I myself do believe in a benevolent God, and I recognize that philosophy and theology may be able to extend the argument.

But a scientific argument for design in biology does not reach that far. Thus while I argue for design, the question of the identity of the designer is left open. Possible candidates for the role of designer include: the God of Christianity; an angel—fallen or not; Plato's demi-urge; some mystical New Age force; space aliens from Alpha Centauri; time travelers; or some utterly unknown intelligent being. Of course, some of these possibilities may seem more plausible than others based on information from fields other than science. Nonetheless, as regards the identity of the designer, modern ID theory happily echoes Isaac Newton's phrase, *hypothesis non fingo.*

Differences From Paley

The fact that modern intelligent design theory is a minimalist argument for design itself, not an argument for the existence of God, relieves it of much of the baggage that weighed down Paley's argument. First of all, it is immune to the argument from evil. It matters not a whit to the scientific case whether the designer is good or bad, interested in us or disinterested. It only matters whether an explanation of design appears to be consistent with the biological examples I point to. Second, questions about whether the designer is omnipotent, or even especially competent, do not arise in my case, as they did in Paley's. Perhaps the designer isn't omnipotent or very competent. More to the point, perhaps the designer was not interested in every detail of biology, as Paley thought, so that while some features were indeed designed, others were left to the vagaries of nature. Thus the modern argument for design need only show that intelligent agency appears to be a good explanation for *some* biological features.

Thus compared to William Paley's argument, modern ID theory

is very restricted in scope. However, what it lacks in scope, it makes up for in resilience. Paley conjoined a number of separable ideas in his argument—design, omnipotence, benevolence, and so on, which made his overall position quite brittle. For example, arguments against the perceived benevolence of the design became arguments against the very existence of design. Thus one got the seeming *non sequitur* stating that because biological feature A appears malevolent, therefore all biological features arose by natural selection or some other unintelligent process.

With the much more modest claims of modern ID theory, such a move is not possible. Attention is kept focused on the basic question of whether unintelligent processes could produce the complex structures of biology, or whether intelligence was indeed required.

Another important point to emphasize right at the beginning is that mine is indeed a *scientific* argument, not a philosophical or theological argument. Let me explain what I mean by that without getting entangled in trying to define those elusive terms. By calling the argument *scientific*, I mean first that it does not rest on any tenet of any particular creed, nor is it a deductive argument from first principles. Rather, it depends critically on physical evidence found in nature. Second, because it depends on physical evidence, it can potentially be falsified by other physical evidence. Thus it is tentative, only claiming that it currently seems to be the best explanation given the information we have available to us right now.

I do acknowledge that the scientific argument for design may have theological implications, but that does not change its status as a scientific idea. I would like to draw a parallel between the modern argument for design in biology and the Big Bang theory in physics. The Big Bang theory strikes many people as having theological implications, as shown by those who do not welcome those implications. For example, in 1989, John Maddox, the editor of *Nature*, the world's leading science journal, published a very peculiar editorial titled "Down with the Big Bang." He wrote:

> Apart from being philosophically unacceptable, the Big Bang is an over-simple view of how the Universe began, and it is unlikely to survive the decade ahead...Creationists ...seeking support for their opinions have ample justification in the Big Bang.[1]

Nonetheless, despite its theological implications, the Big Bang theory is a completely scientific one, which justifies itself by physical data, not by appeals to holy books. I think a theory of intelligent design in biology fits in the same category: While it may have theological implications, it justifies itself by physical data. Furthermore, just as the Big Bang theory could be overturned tomorrow by new evidence, so could intelligent design theory. Both are tentative.

With these preliminary remarks in mind, I turn to considering the scientific case for intelligent design in biology. I will proceed as follows. First, I will briefly make the case for design. Second, I will address several specific scientific objections put forward by critics of design. Finally I will discuss the question of falsifiability.

Darwinism and Design

In 1859 Charles Darwin published his great work *The Origin of Species*, in which he proposed to explain how the great variety and complexity of the natural world might have been produced solely by the action of blind physical processes. His proposed mechanism was, of course, natural selection working on random variation. In a nutshell, Darwin reasoned that the members of a species whose chance variation gave them an edge in the struggle to survive would tend to survive and reproduce. If the variation could be inherited, then over time the characteristics of the species would change. And over great periods of time, perhaps great changes would occur.

It was a very elegant idea. Nonetheless, Darwin knew his proposed mechanism could not explain everything, and in *The Origin of Species* he gave us a criterion by which to judge his theory. He wrote:

> If it could be demonstrated that any complex organ existed

> which could not possibly have been formed by numerous, successive, slight modifications, my theory would absolutely break down.[2]

Adding, however, that he could "find out no such case," Darwin of course was justifiably interested in protecting his fledgling theory from easy dismissal, and so he threw the burden of proof on opponents to prove a negative to "demonstrate" that something "could not possibly" have happened—which is essentially impossible to do in science. Nonetheless let's ask, What might at least *potentially* meet Darwin's criterion? What sort of organ or system seems unlikely to be formed by "numerous, successive, slight modifications"? A good place to start is with one that is *irreducibly complex*. In *Darwin's Black Box: The Biochemical Challenge to Evolution*, I defined an irreducibly complex system as "a single system which is composed of several well-matched, interacting parts that contribute to the basic function, and where the removal of any one of the parts causes the system to effectively cease functioning."[3]

A good illustration of an irreducibly complex system from our everyday world is a simple mechanical mousetrap. A common mousetrap has several parts, including a wooden platform, a spring with extended ends, a hammer, holding bar, and catch. Now, if the mousetrap is missing the spring, or hammer, or platform, it doesn't catch mice half as well as it used to, or a quarter as well. It simply doesn't catch mice at all. Therefore it is irreducibly complex. It turns out that irreducibly complex systems are headaches for Darwinian theory, because they are resistant to being produced in the gradual, step-by-step manner that Darwin envisioned.

As biology has progressed with dazzling speed in the past half century, we have discovered many systems in the cell, at the very foundation of life, which, like a mousetrap, are irreducibly complex. Time permits me to mention only one example here—the bacterial flagellum. The flagellum is quite literally an outboard motor that some bacteria use to swim. It is a rotary device, which, like a motorboat, turns a propeller to push against liquid, moving the bacterium forward in the process. It consists of a number of parts, including a long tail that acts as a propeller, the hook region, which attaches the propeller to the drive shaft, the motor, which uses a flow of acid from the outside of the bacterium to the inside to power the turning, a stator, which keeps the structure stationary in the plane of the membrane while the propeller turns, and bushing material to allow the driveshaft to poke up through the bacterial membrane. In the absence of the hook, or the motor, or the propeller, or the drive shaft, or most of the forty different types of proteins that genetic studies have shown to be necessary for the activity or construction of the flagellum, one doesn't get a flagellum that spins half as fast as it used to, or a quarter as fast. Either the flagellum doesn't work, or it doesn't even get constructed in the cell. Like a mousetrap, the flagellum is irreducibly complex. And like the mousetrap, its evolutionary development by "numerous, successive, slight modifications" is quite difficult to envision. In fact, if one examines the scientific literature, one quickly sees that no one has ever proposed a serious, detailed model for how the flagellum might have arisen in a Darwinian manner, let alone conducted experiments to test such a model. Thus in a flagellum we seem to have a serious candidate to meet Darwin's criterion. We have a system that seems very unlikely to have been produced by "numerous, successive, slight modifications."

Is there an alternative explanation for the origin of the flagellum? I think there is, and it's really pretty easy to see. But in order to see it, we have to do something a bit unusual—we have to break a rule. The rule is rarely stated explicitly. But it was set forth candidly by Nobel laureate Christian DeDuve in his 1995 book *Vital Dust*, in which he speculated on the expansive history of life. He wrote:

> A warning: All through this book I have tried to conform to the overriding rule that life be treated as a natural process, its origin, evolution, and manifestations, up to and including the human species, as governed by the same laws as nonliving processes.[4]

In science journals the rule is always obeyed, at least in letter, yet sometimes it is violated in spirit. For example, several years ago David DeRosier, professor of biology at Brandeis University, published a review article on the bacterial flagellum in which he remarked:

> More so than other motors, the flagellum resembles a machine designed by a human.[5]

That same year the journal *Cell* (February 6, 1998) published a special issue on the topic of "Macromolecular Machines." On the journal's cover was a painting of a stylized protein apparently in the shape of an animal, with what seems to be a watch in the foreground (perhaps Paley's watch). Articles in the journal had titles such as "The Cell As a Collection of Protein Machines," "Polymerases and the Replisome: Machines within Machines," and "Mechanical Devices of the Spliceosome: Motors, Clocks, Springs and Things." By way of introduction, on the contents page was written:

> Like the machines invented by humans to deal efficiently with the macroscopic world, protein assemblies contain highly coordinated moving parts.

Well, if the flagellum and other biochemical systems strike scientists as looking like "machines" that were "designed by a human" or "invented by humans," then why don't we actively entertain the idea that perhaps they were indeed designed by an intelligent being? We don't do that, of course, because it would violate the rule. But sometimes, when a fellow is feeling frisky, he throws caution to the wind and breaks a few rules. In fact, that is just what I did in *Darwin's Black Box*. I proposed that, rather than Darwinian evolution, a more compelling explanation for the irreducibly complex molecular machines discovered in the cell is that they were indeed designed, as David DeRosier and the editors of *Cell* apprehended—purposely designed by an intelligent agent. That proposal has attracted a bit of attention. Some of my critics have asserted that the proposal of intelligent design is a religious idea, not a scientific one. I disagree. I think the conclusion of intelligent design in these cases is completely empirical. That is, it's based entirely on the physical evidence, along with an appreciation for how we come to a conclusion of design. Every day of our lives we decide, consciously or not, that some things were designed, other not. How do we do that? How do we come to a conclusion of design?

To help see how we conclude design, imagine that you are walking with a friend in the woods. Suddenly your friend is pulled up by the ankle by a vine and left dangling in the air. After you cut him down, you reconstruct the situation. You see that the vine was tied to a tree limb that was bent down and held by a stake in the ground. The vine was covered by leaves so that you wouldn't notice it, and so on. From the way the parts were arranged you would quickly conclude that this was no accident—this was a designed trap. That is not a religious conclusion, but one based firmly in the physical evidence.

Although I think that intelligent design is a rather obvious hypothesis, nonetheless my book seems to have caught a number of people by surprise, and so it has been reviewed pretty widely. The *New York Times*, the *Washington Post*, the *Allentown Morning Call*—all the major media have taken a look at it. Unexpectedly, not everyone agreed with me. In fact, in response to my argument, several scientists have pointed to experimental results, which, they claim, either cast much doubt over the claim of intelligent design or outright falsify it. In the remainder of this chapter I will discuss these counterexamples. I hope to show why I think they not only fail to support Darwinism, but why they actually fit much better with a theory of intelligent design. After that, I will discuss the issue of falsifiability.

An "Evolved" Operon

Kenneth Miller, a professor of cell biology at Brown University, has written a book recently titled *Finding Darwin's God*, in which he defends Darwinism from a variety of critics, including myself. In a chapter devoted to rebutting *Darwin's Black Box*, he quite correctly states that "a true acid test" of the ability of Darwinism to deal with irreducible complexity would be to "[use] the tools of molecular genetics to wipe out an existing multipart system and then see if evolution can come to the rescue with a system to replace it."[6] He then cites the careful work over the past twenty-five years of Barry Hall of the University of Rochester on the experimental evolution of a lactose-utilizing system in *E. coli*.

Here is a brief description of how the system, called the *lac operon*, functions. The lac operon of *E. coli* contains genes coding for several proteins that are involved in the metabolism of a type of sugar called *lactose*. One protein of the lac operon, called a *permease*, imports lactose through the otherwise impermeable cell membrane. Another protein is an enzyme called *galactosidase*, which can break down lactose to its two constituent monosaccharides, galactose and glucose, which the cell can then process further. Because lactose is rarely available in the environment, the bacterial cell switches off the genes until lactose is available. The switch is controlled by another protein called a *repressor*, whose gene is located next to the operon. Ordinarily the repressor binds to the lac operon, shutting it off by physically interfering with the operon. However, in the presence of the natural "inducer" allolactose or the artificial chemical inducer IPTG, the repressor binds to the inducer and releases the operon, allowing the lac operon enzymes to be synthesized by the cell.

After giving his interpretation of Barry Hall's experiments, Kenneth Miller excitedly remarks:

> Think for a moment—if we were to happen upon the interlocking biochemical complexity of the reevolved lactose system, wouldn't we be impressed by the intelligence of its design? Lactose triggers a regulatory sequence that switches on the synthesis of an enzyme that then metabolizes lactose itself. The products of that successful lactose

> metabolism then activate the gene for the lac permease, which ensures a steady supply of lactose entering the cell. Irreducible complexity. What good would the permease be without the galactosidase? No good, of course.
>
> By the very same logic applied by Michael Behe to other systems, therefore, we could conclude that the system had been designed. Except we *know* that it was *not* designed. We know it evolved because we watched it happen right in the laboratory! No doubt about it—the evolution of biochemical systems, even complex multipart ones, is explicable in terms of evolution. Behe is wrong.[7]

For the next few minutes I will try to show that the picture Miller paints is greatly exaggerated. In fact, far from being a difficulty for design, the very same work that Miller points to as an example of Darwinian prowess I would cite as showing the limits of Darwinism and the need for design.

So what did Barry Hall actually do? To study bacterial evolution in the laboratory, in the mid-1970s Hall produced a strain of *E. coli* in which the gene for *just* the galactosidase of the lac operon was deleted. He later wrote:

> All of the other functions for lactose metabolism, including lactose permease and the pathways for metabolism of glucose and galactose, the products of lactose hydrolysis, remain intact, thus re-acquisition of lactose utilization requires only the evolution of a new ß-galactosidase function.[8]

Thus, contrary to Miller's own criterion for "a true acid test," a multipart system was not "wiped out"—only one component of a multipart system was deleted. The lac permease remained intact. What's more, as we shall see, the artificial inducer IPTG was added to the bacterial culture, and an alternate, cryptic galactosidase was left intact.

Without galactosidase, Hall's cells could not grow when cultured on a medium containing only lactose as a food source.

However, when grown on a plate that also included alternative nutrients, bacterial colonies could be established. When the other nutrients were exhausted the colonies stopped growing. However, Hall noticed that after several days to several weeks, hyphae grew on some of the colonies. Upon isolating cells from the hyphae, Hall saw that they frequently had two mutations, one of which was in a gene for a protein he called "evolved ß-galactosidase" (ebg), which allowed it to metabolize lactose efficiently. The ebg gene is located in another operon, distant from the lac operon, and is under the control of its own repressor protein. The second mutation Hall found was always in the gene for the ebg repressor protein, which caused the repressor to bind lactose with sufficient strength to de-repress the ebg operon.

The fact that there were two separate mutations in different genes—neither of which by itself allowed cell growth[9]—startled Hall, who knew that the odds against the mutations appearing randomly and independently were prohibitive.[10] Hall's results and similar results from other laboratories led to research in a new area dubbed "adaptive mutations."[11] As Hall later wrote:

> Adaptive mutations are mutations that occur in nondividing or slowly dividing cells during prolonged nonlethal selection, and that appear to be specific to the challenge of the selection in the sense that the only mutations that arise are those that provide a growth advantage to the cell. The issue of the specificity has been controversial because it violates our most basic assumptions about the randomness of mutations with respect to their effect on the cell.[12]

The mechanisms of adaptive mutation are currently unknown. While they are being sorted out, it seems unwise to cite results of processes that "violate our most basic assumptions about the randomness of mutations" to argue for Darwinian evolution, as Miller does.

The nature of adaptive mutation aside, a strong reason to consider Barry Hall's results to be quite modest is that the ebg proteins—both the repressor and galactosidase—are homologous to the *E. coli* lac proteins and overlap the proteins in activity. Both of the unmutated ebg proteins already bind lactose. Binding of lactose even to the unmutated ebg repressor induces a one-hundred-fold increase in synthesis of the ebg operon.[13] Even the unmutated ebg galactosidase can hydrolyze lactose at a level of about 10 percent of a "Class II" mutant galactosidase that supports cell growth.[14] These activities are not sufficient to permit growth of *E. coli* on lactose, but they already are present. The mutations reported by Hall simply enhance preexisting activities of the proteins.

In a recent paper Professor Hall pointed out that both the lac and ebg galactosidase enzymes are part of a family of highly conserved galactosidases, identical at thirteen of fifteen active sites of amino acid residues, which apparently diverged by gene duplication more than two billion years ago.[15] The two mutations in ebg galactosidase that increase its ability to hydrolyze lactose change two nonidentical residues back to those of other galactosidases, so that their active sites are identical. Thus—before any experiments were done—the ebg active site was already a near duplicate of other galactosidases, and only became more active by becoming a complete duplicate. Significantly, by phylogenetic analysis Hall concluded that those two mutations are the *only* ones in *E. coli* that confer the ability to hydrolyze lactose—that is, no other protein, no other mutation in *E. coli* will work. Hall wrote:

> The phylogenetic evidence indicates that either Asp-92 and Cys/Trp 977 are the only acceptable amino acids at those positions, or that all of the single base substitutions that might be on the pathway to other amino acid replacements at those sites are so deleterious that they constitute a deep selective valley that has not been traversed in the 2 billion years since those proteins diverged from a common ancestor.[16]

To my mind, such results hardly support extravagant claims for the creativeness of Darwinian processes.

Another critical caveat not mentioned by Kenneth Miller is that the mutants that were initially isolated would be unable to use lactose in the wild—they required the artificial inducer IPTG to be present in the growth medium. As Barry Hall states clearly, in the absence of IPTG, no viable mutants are seen. The reason is that a permease is required to bring lactose into the cell. However, ebg only has a galactosidase activity, not a permease activity, so the experimental system had to rely on the preexisting lac permease. Since the lac operon is repressed in the absence of either allolactose or IPTG, Hall decided to include the artificial inducer in all media up to this point so that the cells could grow. Thus the system was being artificially supported by intelligent intervention.

The prose in Miller's book obscures the facts that most of the lactose system was already in place when the experiments began, that the system was carried through nonviable states by inclusion of IPTG, and that the system will not function without preexisting components. From a skeptical perspective, the admirably careful work of Barry Hall involved a series of micromutations stitched together by intelligent intervention. He showed that the activity of a deleted enzyme could be replaced only by mutations to a second, homologous protein with a nearly identical active site, and only if the second repressor already bound lactose, and only if the system were also artificially induced by IPTG, and only if the system were also allowed to use a preexisting permease. In my view, such results are entirely in line with the expectations of irreducible complexity requiring intelligent intervention, and of limited capabilities for Darwinian processes.

Blood Clotting

A second putative counterexample to intelligent design concerns the blood-clotting system. Blood clotting is a very intricate biochemical process, requiring many protein parts. I had devoted a chapter of *Darwin's Black Box* to the blood-clotting cascade, claiming that it is irreducibly complex and so does not fit well within a Darwinian framework. However, Russell Doolittle, a prominent biochemist, member of the National Academy of Sciences, and expert on blood clotting, disagreed. While discussing the similarity of the proteins of the blood-clotting cascade to each other in an essay in *Boston Review* in 1997, he remarked that "the genes for new proteins come from the genes for old ones by gene duplication."[17] Doolittle's invocation of gene duplication has been repeated by many scientists reviewing my book, but it reflects a common confusion. Genes with similar sequences only suggest common descent—they do not speak to the mechanism of evolution. This point is critical to my argument and bears emphasis—*evidence of common descent is not evidence of natural selection.* Similarities among either organisms or proteins are the evidence for descent with modification, that is, for evolution. Natural selection, however, is a proposed explanation for how evolution might take place—its mechanism—and so it must be supported by other evidence if the question is not to be begged.

Doolittle then cited a paper titled "Loss of fibrinogen rescues mice from the pleiotropic effects of plasminogen deficiency."[18] (By way of explanation, *fibrinogen* is the precursor of the clot material; *plasminogen* is a protein that degrades blood clots.) He commented:

> Recently the gene for plaminogen [sic] was knocked out of mice, and, predictably, those mice had thrombotic complications because fibrin clots could not be cleared away. Not long after that, the same workers knocked out the gene for fibrinogen in another line of mice. Again, predictably, these mice were ailing, although in this case hemorrhage was the problem. And what do you think happened when these two lines of mice were crossed? For all practical purposes, the mice lacking both genes were normal! Contrary to claims about irreducible complexity, the entire ensemble of proteins is not needed. Music and harmony can arise from a smaller orchestra.[19]

The implied argument seems to be that the modern clotting

system is actually not irreducibly complex, so a simpler clotting cascade might be missing factors such as plasminogen and fibrinogen, and perhaps it could be expanded into the modern clotting system by gene duplication. However, that interpretation does not stand up to a careful reading of Bugge et al.

In their paper, Bugge et al. note that the lack of plasminogen in mice results in many problems, such as high mortality, ulcers, severe thrombosis, and delayed wound healing. On the other hand, lack of fibrinogen results in failure to clot, frequent hemorrhage, and death of females during pregnancy. The point of Bugge et al. was that if one crosses the two knockout strains, producing plasminogen-plus-fibrinogen deficiency in individual mice, the mice do not suffer the many problems that afflict mice lacking plasminogen alone. Since the title of the paper emphasized that mice are "rescued" from some ill effects, one might be misled into thinking that the double-knockout mice were normal. They are not. As Bugge et al. state in their abstract, "Mice deficient in plasminogen and fibrinogen are phenotypically indistinguishable from fibrinogen-deficient mice." In other words, the double-knockouts have all the problems that mice lacking only fibrinogen have: They do not form clots, they hemorrhage, and the females die if they become pregnant. They are definitely not promising evolutionary intermediates.

The probable explanation is straightforward. The pathological symptoms of mice missing just plasminogen apparently are caused by uncleared clots. But fibrinogen-deficient mice cannot form clots in the first place. So problems due to uncleared clots don't arise either in fibrinogen-deficient mice or in mice that lack both plasminogen and fibrinogen. Nonetheless, the severe problems that attend lack of clotting in fibrinogen-deficient mice continue in the double knockouts. Pregnant females still perish.

Most important for the issue of irreducible complexity, however, is that the double-knockout mice do not merely have a less sophisticated but still functional clotting system. They have no functional clotting system at all. They are not evidence for the Darwinian evolution of blood clotting. Therefore my argument, that the system is irreducibly complex, is unaffected by that example.

Other work from the same laboratory is consistent with the view that the blood-clotting cascade is irreducibly complex. Experiments with "knockout" mice in which the genes for other clotting components, called *tissue factor* and *prothrombin*, have been deleted separately show that those components are required for clotting, and in their absence the organism suffers severely.[20]

In ending this section, let me just make explicit the point that two very competent scientists, Professors Miller and Doolittle, both of whom are highly motivated to discredit claims of intelligent design, and both of whom are quite capable of surveying the entire biomolecular literature for experimental counterexamples, came up with examples that, when looked at skeptically, actually buttress the case for irreducible complexity rather than weaken it. Of course this does not prove that claims of irreducible complexity are true or that intelligent design is correct. But it does show, I think, that scientists don't have a handle on irreducible complexity, and that the idea of intelligent design is considerably stronger than its detractors would have us believe. It also shows the need to treat Darwinian scenarios, such as Miller and Doolittle offered, with a hermeneutic of suspicion. Some scientists believe so strongly in Darwinism that their critical judgments are affected, and they will unconsciously overlook pretty obvious problems with Darwinian scenarios or confidently assert things that are objectively untrue.

Falsifiability

Let us now consider the issue of falsifiability. Let me say up front that I know most philosophers of science do not regard falsifiability as a necessary trait of a successful scientific theory. Nonetheless, falsifiability is still an important factor to consider since it is nice to know whether or not one's theory can be shown to be wrong by contact with the real world.

A frequent charge made against intelligent design is that it is unfalsifiable, or untestable. For example, in its recent booklet *Science and Creationism* the National Academy of Sciences writes:

> [I]ntelligent design…[is] not science because [it is] not testable by the methods of science.[21]

Yet that claim seems to be at odds with the criticisms I have just summarized. Clearly, Russell Doolittle and Kenneth Miller advanced scientific arguments aimed at falsifying intelligent design. If the results of Bugge et al. had been as Doolittle first thought, or if Barry Hall's work had indeed shown what Miller implied, then they correctly believed my claims about irreducible complexity would have suffered quite a blow.

Now, one can't have it both ways. One can't say both that intelligent design is unfalsifiable (or untestable) and that there is evidence against it. Either it is unfalsifiable and floats serenely beyond experimental reproach, or it can be criticized on the basis of our observations and is therefore testable. The fact that critical reviewers advance scientific arguments against intelligent design (whether successfully or not) shows that intelligent design is indeed falsifiable. What's more, it is wide open to falsification by a series of rather straightforward laboratory experiments such as those that Miller and Doolittle pointed to, which is exactly why they pointed to them.

Now let's turn the tables and ask, how could one falsify the claim that a particular biochemical system was produced by a Darwinian process? Kenneth Miller announced an "acid test" for the ability of natural selection to produce irreducible complexity. He then decided that the test was passed and unhesitatingly proclaimed intelligent design to be falsified ("Behe is wrong"[22]). But if, as it certainly seems to me, *E. coli* actually fails the lactose-system "acid test," would Miller consider Darwinism to be falsified? Almost certainly not. He would surely say that Barry Hall started with the wrong bacterial species, or used the wrong selective pressure, and so on. So it turns out that his "acid test" was not a test of Darwinism; it tested only intelligent design.

The same one-way testing was employed by Russell Doolittle. He pointed to the results of Bugge et al., saying they argue against intelligent design. But when the results turned out to be the opposite of what he had originally thought, Professor Doolittle did not abandon Darwinism.

It seems then, perhaps counterintuitively to some, that intelligent design is quite susceptible to falsification, at least on the points under discussion. Darwinism, on the other hand, seems quite impervious to falsification. The reason for that can be seen when we examine the basic claims of the two ideas with regard to a particular biochemical system like, say, the bacterial flagellum. The claim of intelligent design is that "*no* unintelligent process could produce this system." The claim of Darwinism is that "*some* unintelligent process could produce this system." To falsify the first claim, one need only show that at least one unintelligent process could produce the system. To falsify the second claim, one would have to show the system could not have been formed by any of a potentially infinite number of possible unintelligent processes, which is effectively impossible to do.

The danger of accepting an effectively unfalsifiable hypothesis is that science has no way to determine if the belief corresponds to reality. In the history of science, the scientific community has believed in any number of things that were in fact not true, not real—for example, the universal ether. If there were no way to test those beliefs, the progress of science might be substantially and negatively affected. If, in the present case, the expansive claims of Darwinism are, in reality, not true, then its unfalsifiability will cause science to bog down in these areas, as I believe it has.

So, what can be done? I don't think that the answer is to never investigate a theory that is unfalsifiable. After all, although it is unfalsifiable, Darwinism's claims are potentially positively demonstrable. For example, if some scientist conducted an experiment showing the production of a flagellum (or some equally complex system) by Darwinian processes, then the Darwinian claim would be affirmed. The question only arises in the face of negative results.

I think several steps can be prescribed. First of all, one has to be aware—raise one's consciousness—about when a theory is unfalsifiable. Second, as far as possible, an advocate of an unfalsifiable theory should try as diligently as possible to demonstrate positively the claims of the hypothesis. Third, one needs to relax Darwin's criterion from this:

> If it could be demonstrated that any complex organ existed which could not possibly have been formed by numerous, successive, slight modifications, my theory would absolutely break down.[23]

...to something like this:

> If a complex organ exists that seems very unlikely to have been produced by numerous, successive, slight modifications, and if no experiments have shown that it or comparable structures can be so produced, then maybe we're *barking up the wrong tree. So let's break some rules!*

Of course, people will differ on the point at which they decide to break rules. But at least with the realistic criterion there could be evidence against the unfalsifiable. At least then people like Doolittle and Miller would run a risk when they cite an experiment that shows the opposite of what they had thought. At least then science would have a way to escape from the rut of unfalsifiability and think new thoughts.*

*"The Modern Intelligent Design Hypothesis" was first published in *Philosophia Christi* Series 2, vol. 3, no. 1 (2001): 165–179. It is reprinted here by permission of the author, Dr. Michael Behe.

Michael J. Behe, Ph.D.

Michael J. Behe, Ph.D. needs no introduction, especially in light of his prominence in the previous chapters. His work on irreducibly complex systems inside the cell has shaken the foundations of evolutionary biology and has generated a tidal wave of published responses. In this chapter, Dr. Behe recaps his critique of Darwinism and proceeds to outline the major responses he has received. Then, one by one, he shows why these criticisms do not even begin to solve the problem of irreducible complexity. In fact, they validate his criticism and highlight even further the severity of the problem that presents itself to neo-Darwinism.

Dr. Behe approaches the hypothesis of *intelligent design* from a strictly scientific point of view. Thus he shows why—since the identity of the creator cannot be pinned down by the tools of science—intelligent design theory serves as a practical and fully appropriate approach, even in a secular scientific arena.

Now that the black box of vision has been opened, it is no longer enough for an evolutionary explanation of that power to consider only the *anatomical* structures of whole eyes, as Darwin did in the nineteenth century (and as popularizers of evolution continue to do today). Each of the anatomical steps and structures that Darwin thought were so simple actually involves staggeringly complicated biochemical processes that cannot be papered over with rhetoric.

—Michael J. Behe
From "Lilliputian Biology" in *Darwin's Black Box*

On the Design of the Vertebrate Retina

By George Ayoub, Ph.D.
Department of Biology, Westmont College
Santa Barbara, CA

It has been commonly claimed that the vertebrate eye is functionally suboptimal, because photoreceptors in the retina are oriented away from incoming light. However, there are excellent functional reasons for vertebrate photoreceptors to be oriented as they are. Photoreceptor structure and function is maintained by a critical tissue, the retinal pigment epithelium (RPE), which recycles photopigments, removes spent outer segments of the photoreceptors, provides an opaque layer to absorb excess light, and performs additional functions. These aspects of the structure and function of the vertebrate eye have been ignored in evolutionary arguments about suboptimality, yet they are essential for understanding how the eye works.

Editorial Introduction: A Popular Argument

It has been widely argued in both the technical (Thwaites 1984,[1] Williams 1992[2]) and popular evolutionary literature (Diamond 1985,[3] Dawkins 1986,[4] Miller 1994[5]) that the vertebrate eye is poorly designed. "In fact it is stupidly designed," writes the influential neo-Darwinian theorist George Williams, "because it embodies many functionally arbitrary or maladaptive features."[6] Chief among these features, Williams claims, is the *inversion of the retina.*[7]

"The retina is upside down," he writes. "The rods and cones are the bottom layer, and light reaches them only after passing through the nerves and blood vessels." These structures, claims UCLA evolutionary biologist Jared Diamond, aren't located behind the photoreceptors, where any sensible engineer would have put them, but out in front of them, where they screen some of the incoming light.[8] A camera designer who committed such a blunder would be fired immediately.

The capstone of this argument is held to be the cephalopod (squid and octopus) retina, which is putatively "wired correctly," with its photoreceptors facing *toward* the light and with its nerves "neatly tucked away behind the photoreceptor layer."[9] The cephalopods, it is said, got it right.

In considering this argument, we may dispense immediately with optimality comparisons between cephalopod (invertebrate) and vertebrate retina designs. None of the authors cited above provide any evidence that the cephalopod retina is functionally superior to the vertebrate retina: a claim that, in any case, seems unreasonable on its face. Would hundreds of thousands of vertebrate species—in a great variety of terrestrial, marine, and aerial environments—really see better with a visual system used by a handful of exclusively marine vertebrates? In the absence of any rigorous comparative evidence, all claims that the cephalopod retina is functionally superior to the vertebrate retina remain entirely conjectural. In short, there is no reason to believe them.

But we should consider a more basic point. Why refer to the cephalopod retina at all? The claim that the cephalapods got it right assumes that the vertebrates did not, and that the latter are making the best of a bad situation—but, of course, it remains to be demonstrated that, *in fact*, the vertebrate retina is suboptimal.

And this has not been demonstrated—not by the authors cited above or by other evolutionary biologists. "One of the difficulties with the hypothesis of optimality," note Farnsworth and Niklas, "is the availability of observations to test it."[10] That goes as well for hypotheses of suboptimality, as exemplified by the evolutionary literature on the vertebrate retina. The biological world is full of puzzling systems. While it is not readily apparent why vertebrate photoreceptors face away from the light, nor why other cell layers intervene, a good many things in science are not apparent at first glance. We need to look more deeply.

In this case, we need not look far. There are excellent functional reasons for vertebrate photoreceptors to be oriented as they are. These aspects of retinal structure and function have been ignored in evolutionary arguments about suboptimality, yet they are essential for understanding how the eye works.

The Structure of the Vertebrate Retina

First, some anatomy. Figure 1 depicts a vertebrate eye in cross section. (See page 154.) Light passes through the cornea, the primary focusing element, then through the iris, which controls how much light will enter the eye, and lastly, through the lens, which provides the adjustable focusing element. The light, now adjusted for intensity, is focused onto the thin tissue lining the back of the eye: the retina.

The retina (see Figure 2, page 156) comprises cells from the central nervous system (CNS) and converts, or transduces light into electrical signals, the "medium" of the CNS. A highly complex tissue, the retina contains cells of several different types:

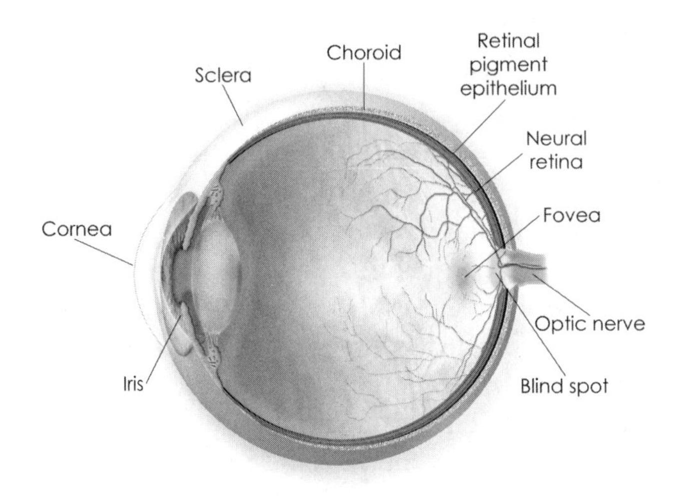

Figure 1—Eye Anatomy

- *Photoreceptors* (rods and cones, Figure 2) actually convert the light energy into an electrical signal. These are the first cells directly involved in communicating information within the visual system, sending signals via a chemical synapse to...
- *The bipolar cells* (Figure 2), the second order cells in the retina; the bipolar cells synapse onto...
- *The ganglion cells* (Figure 2), at the inner surface of the retina; these cells have axons that travel together from the eye, exiting via the optic disk and forming the optic nerve en route to the brain.
- *The amacrine cells* (Figure 2) mediate lateral interactions, transmitting information between adjacent

bipolar cells and ganglion cells and the horizontal cells that communicate laterally in the outer retina.

The light must pass first through all the auxiliary cells before arriving at the photoreceptors—which at first glance hardly seems sensible. If the design problem to be solved by any eye is forming a maximally accurate image of the world, then degrading the light before it reaches the "business end" of the photoreceptors seems self-evidently a poor solution. "This is equivalent to placing a thin diffusing screen directly over the film in your camera; it can only degrade the quality of the image."[11]

And that is where evolutionary accounts leave the story.

The Critical Role of the Retinal Pigmented Epithelium (RPE)

But there is much more to be said. Lying directly behind the retina is an epithelial tissue that maintains the photoreceptors. (See Figure 2, page 156.) This tissue, called the retinal pigmented epithelium (hereafter, RPE), is critical to the development and function of the retina. Indeed, volumes have been dedicated to understanding the role of the RPE, because when it malfunctions, the eye as a whole malfunctions.[12]

1. Regenerating photoreceptive pigments

When light strikes a photoreceptor, it sets in motion a chain of molecular events that eventually culminate in forming an image in the brain. Here, let's focus on just the first characters in the story.

The first player on stage is the photosensitive molecule rhodopsin. Rhodopsin consists of a protein, opsin, and another molecule, 11-cis-retinal. Found at the distal ends of the photoreceptors—the portion closest to the RPE, called the outer segment—rhodopsin is embedded in membranous discs. (An important note about some potentially confusing terminology: The "business end" of the photoreceptor cell, where the membranous discs occur, is called the outer segment. In vertebrates, however, this segment is actually inner, that is, at the back of the retina, pointing in toward the center of the organism.) When light strikes rhodopsin, the energy changes the shape of its molecular component, 11-cis-retinal, into an all-trans conformation, a process called isomerization. This conformational change in retinal starts a complex cascade of reactions in several other molecules, causing the hyperpolarization (or shift in electrical charge) of the outer segment membrane. Molecular transmitters then carry this electrical signal from the synapse at the photoreceptor's base to the next neurons, the horizontal cells and bipolar cells—thus beginning the process by which we see.

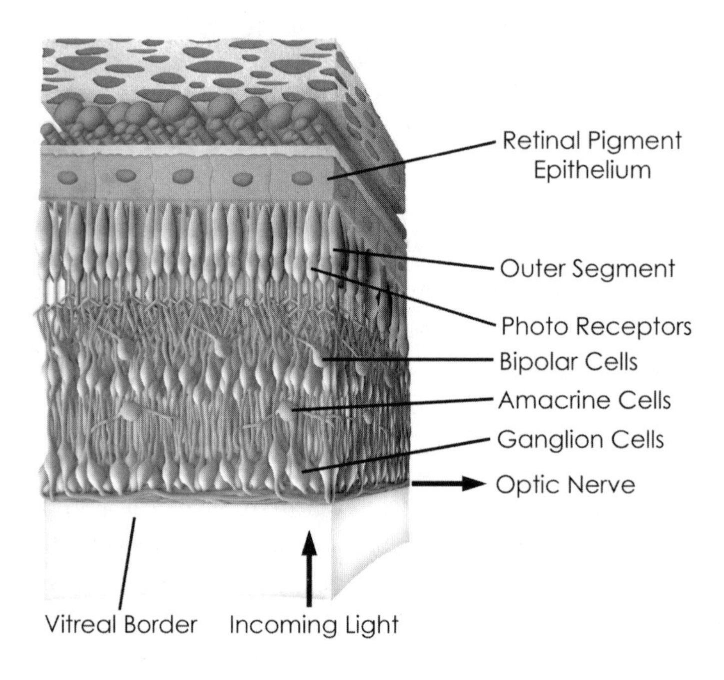

Figure 2—Retinal Layers

This process depends critically on the isomerization of 11-cis-retinal. Each photon of light striking a photoreceptor can isomerize retinal, and since many billions of photons constantly strike the eye, retinal must be replaced regularly to maintain the cycle and overall photoreceptor function. That job of replacement falls to the RPE. The RPE cells collect the used retinal from the photoreceptors and employ vitamin A to make fresh retinal, transporting it back to the photoreceptors.[13]

2. *Recycling of photoreceptive material*

Next on the list of RPE responsibilities is a related function—recycling the used outer segments. Outer segment membranes are very active and thus must be continually replaced.

Each day, new outer segment membrane grows at the base of the outer segment (where it intersects with the inner segment, the cell region containing the nucleus), adding to the length of the photoreceptor. As the outer segment lengthens from its base, its distal end—the oldest membrane—sheds in segments. These segments are picked up by the RPE, which phagocytizes the material, recycling all of the molecules present.[14]

Thus, spent photoreceptive membranes are removed from the optical path, to be replaced by new material. This process, which goes on continually, maintains the high sensitivity of the photoreceptors.[15]

3. *Absorption of excess light*

In addition to these active functions, the RPE also has an important passive role. Because it is heavily pigmented, it forms an opaque screen behind the optical path of the photoreceptors.

It thus absorbs light that is not collected by the photoreceptors, light that would otherwise decrease the resolution of images. This absorptive property of the RPE is important to maintaining high visual acuity.

The RPE "is required for the normal development of the eye," a function that, while not directly related to vision, certainly undergirds the very possibility of seeing at all.[16] In short:

> Considering the diverse functions of the RPE cells...there is no doubt that the integrity of the RPE metabolic machinery is essential for the normal functioning of the outer retina. Because of the nature of these interactions, it is essential that the RPE and photoreceptors be in close proximity for normal retinal function.[17]

There are excellent reasons for vertebrate photoreceptors to be oriented as they are.

A Thought Experiment

But still—there sits a blind spot in each retina. To be sure, the blind spots are displaced laterally from each other, so that "with both eyes open, we can see everything in the visual field" as one eye sees what the other does not.[18] However, we can imagine situations where this wouldn't work:

> Our retinal blind spots rarely cause any difficulty, but *rarely* is not the same as *never*. As I momentarily cover one eye to ward off an insect, an important event might be focused on the blind spot of the other.[19]

So, as a thought experiment, let's fix the blind spot. We will start by turning the photoreceptors around, so their wiring isn't in the way.

We have eliminated the blind spot, providing slightly better sight in one portion of the eye. Now, however, the blood vessels and RPE, needed to maintain the photoreceptors, must be located on the *inner* side of the retina, between it and the lens. This places a large capillary bed (containing many red blood cells) and an epithelial tissue in the path of the light, significantly degrading the visual information passing to the photoreceptors.

Furthermore, since the photoreceptors continually shed material from their outer segments, dumping this opaque waste in the path of the light would greatly diminish the amount of light reaching the photoreceptors. Our proposed change also reduces the quality of the light by refracting it with the opaque pieces of shed outer segment membrane.

We might imagine simply placing the RPE at the back of the retina, but this raises the problem of how to dispose of spent outer segment membranes so that the photoreceptors can be quickly regenerated. Or, perhaps, we could surround each photoreceptor cell by RPE cells, but this would increase the space between the photoreceptors, thus decreasing the resolution of vision.

These design changes may force temporal or spatial decrements in vision.

Are these improvements? Hardly; indeed, our thought experiment has taken the vertebrate eye rapidly downhill. In trying to eliminate the blind spot, we have generated a host of new and more severe functional problems to solve. Our "repair" seems far worse than the apparent flaw we wanted to fix.

Conclusion

The vertebrate retina provides an excellent example of functional—though nonintuitive—design. The design of the retina is responsible for its high acuity and sensitivity. It is simply untrue that the retina is demonstrably suboptimal, nor is it easy to conceive how it might be modified without significantly decreasing its function.*

*"On the Design of the Vertebrate Retina" was first published in *Origins and Design* 17:1 (1996). It is reprinted here by permission of the author, Dr. George Ayoub.

George Ayoub

George Ayoub, Ph.D., who teaches biology at Westmont College in Santa Barbara, California, received his Ph.D. in biology at Baylor University School of Medicine. He has done extensive research on the morphology and physiology of the vertebrate retina. This chapter details the evidence for intelligent design of the unique structure of this retina, in response to the common charge from evolutionists that the retina is "wired backward," thus betraying a clumsy evolutionary origin.

Here is, perhaps, the most dramatic example of the principle that wherever we find significant empirical discontinuities in nature we invariably face great, if not insurmountable, conceptual problems in envisaging how the gaps could have been bridged in terms of gradual random processes. We saw this in the fossil record, we saw it in the case of the feather, in the case of the avian lung and in the case of the wing of the bat. We saw it again in the case of the origin of life, and we see it here in this new area of comparative biochemistry.

—Michael Denton

From "A Biochemical Echo of Typology" in *Evolution: A Theory in Crisis*

TEN

Chapter 10

The Mystery of the Yeast Genome

By Thomas E. Woodward, Ph.D.

Scientists recently cracked open the genetic library of the humble creature we call yeast and read it from beginning to end. They expected that this library would reveal telltale marks of its supposed origin from a first cousin, bacteria. *What they found on yeast's DNA bookshelves shocked the scientists. The results seem to throw yet another monkey wrench into Darwin's theory.*

The object of scrutiny that produced such a shock was incredibly tiny. It is the "complete library of DNA" found in each cell of baker's yeast—a genetic microlibrary that scientists call the *genome.* This spool of genetic tape was recently opened up using clever chemical probes and was read by scientists. Right down to the last DNA paragraph—the very last genetic word in baker's yeast—this unique string of information has finally been completely deciphered and written down, letter by letter. Scientists call this process *sequencing* the DNA.[1]

Guess what? The preliminary analyses don't look good for Darwinian evolution. The most shocking feature of the yeast genome is the huge number of so-called *orphan genes.* These are genes that seem to have no *parents* in supposedly older organisms.

Consider the big picture of this tiny *genome* library. Each yeast cell genome has about 6,300 genes. Each gene is a unique segment, or string, of DNA that has several hundred chemical letters arranged in linear fashion, as if assembled on a huge Scrabble board. Yet 3,000 of these genes are of unknown function and have no genetic parallels (called *homologues*) among the other kinds of living things. Neither the so-called *archaebacteria* (weird strains that love extreme environments like boiling water) nor the ordinary *eubacteria*, or any other known animals (human, fruit fly, and so on), have similar genes. This is great news for geneticists and biochemists because it means humble baker's yeast is loaded with untapped material. To flesh out a picture of this discovery, let's use the analogy of a library. Think of each gene as a short paperback book. It appears that approximately 50 percent of yeast's DNA library is filled with rare books that are truly exotic—they have never been seen before in any other living thing. The genome of yeast is uniquely specified for life *as baker's yeast* (and not simply life as a member of the animal kingdom).

But even among the *50 percent* of yeast genes that have parallels in other animals, things look troublesome for our evolutionary friends. In the May 29, 1997, issue of *Nature*, Clayton and his fellow researchers summarize the findings and present an initial comparative analysis among all genomes sequenced. They took the sequence similarities (ignoring the orphan genes) and clustered them into *orthologue groups*. They define such groups as follows: "groups of genes from different organisms which have the same function."[2]

As far as evolutionary relationships go, this is very troublesome. Note the tone of worry in this summary presentation:

> With such a small set of common genes, and so many unique genes, we are led to question the concept of *S. cerevisiae* and *E. coli* as model organisms—that is, physiological representatives of many other organisms. In a real sense, *organisms that contain unique constituent parts* cannot be models for another. Having the whole genomes spread out for our examination *underscores how different from one another these organisms are.*[3]

See the trouble for evolution? These authors suggest that these organisms are so different from each other, it is hard to justify using one to explain the workings of another. *Now, are these organisms atypical?* There are two reasons for thinking not. First, are we really so unlucky that we happened to choose independently two systems that are too unique? Second and more important, we have four very different eubacterial sequences in front of us. And there is no reason to think *E. coli* is a worse "model" than any of the others.

Of course, maybe it shouldn't surprise us "how different from one another these organisms are." For years, we have heard about claims of similarities between organisms. But anyone who knows the molecular biology and genetics behind these studies should realize these uncovered similarities are probably due to sampling bias. You see, it's been very helpful to study a process that has an analog in another better-characterized organism. Thus, much of science in the last two decades has converged on the similarities simply because it makes the *work* much easier (which, of course, translates as a better record of publications).

Thus, these published genomes might be telling us not that yeast and *E. coli* are bad "model organisms," but that we have been overlooking the profound differences because they don't fit into our research paradigms (thus, no true *model organism* exists). It will be very interesting to see genomic comparisons of different phyla and classes (although none of us will probably live that long).*

*The author would like to thank Dr. Mike Gene for his extensive research into the "yeast genome mystery" and for his help with an earlier version of this chapter.

Unlike design elements of the past, the claim that transcendent design pervades the universe is no longer a strictly philosophical or theological claim. It is also a fully scientific claim and follows directly from the complexity-specification criterion. In particular, this is not an argument from ignorance...Natural causes are too stupid to keep pace with intelligent causes. We've suspected this all along. Intelligent design theory provides a rigorous scientific demonstration of this long-standing intuition.

—William Dembski
From "The Act of Creation" *in Intelligent Design*

ELEVEN

Chapter 11

Science and Design

BY WILLIAM A. DEMBSKI, PH.D.

When the physics of Galileo and Newton displaced the physics of Aristotle, scientists tried to explain the world by discovering its deterministic natural laws. When the quantum physics of Bohr and Heisenberg in turn displaced the physics of Galileo and Newton, scientists realized they needed to supplement their deterministic natural laws by taking into account chance processes in their explanations of our universe. *Chance and necessity*, to use a phrase made famous by Jacques Monod, thus set the boundaries of scientific explanation.

Today, however, chance and necessity have proven insufficient to account for all scientific phenomena. Without invoking the rightly discarded teleologies, entelechies, and vitalisms of the past, one can still see that a third mode of explanation is required, namely, intelligent design. Chance, necessity, and design—these three modes of explanation—are needed to explain the full range of scientific phenomena.

Not all scientists see that excluding intelligent design artificially restricts science, however. Richard Dawkins, an arch-Darwinist, begins his book *The Blind Watchmaker* by stating, "Biology is the study of complicated things that give the appearance of having been designed for a purpose."[1] Statements like this echo throughout the biological literature. In *What Mad Pursuit*, Francis Crick, Nobel laureate and codiscoverer of the structure of DNA, writes, "Biologists must constantly keep in mind that what they see was not designed, but rather evolved."[2]

The biological community thinks it has accounted for the apparent design in nature through the Darwinian mechanism of random mutation and natural selection. The point to appreciate, however, is that in accounting for the apparent design in nature, biologists regard themselves as having made a successful *scientific* argument against actual design. This is important, because for a claim to be scientifically falsifiable, it must have the possibility of being true. Scientific refutation is a double-edged sword. Claims that are refuted scientifically may be wrong, but they are not *necessarily* wrong—they cannot simply be dismissed out of hand.

To see this, consider what would happen if microscopic examination revealed that every cell was inscribed with the phrase "Made by Yahweh." Of course, cells don't have "Made by Yahweh" inscribed on them, but that's not the point. The point is that we wouldn't know this unless we actually looked at cells under the microscope. And if they were so inscribed, one would have to entertain the thought, as a scientist, that they actually were made by Yahweh. So even those who do not believe in it tacitly admit that design always remains a live option in biology. *A priori* prohibitions against design are philosophically unsophisticated and easily countered. Nonetheless, once we admit that design cannot be excluded from science without argument, a weightier question remains: Why should we want to admit design into science?

To answer this question, let us turn it around and ask instead, Why shouldn't we want to admit design into science? What's wrong with explaining something as designed by an intelligent agent? Certainly there are many everyday occurrences that we explain by appealing to design. Moreover, in our workaday lives it is absolutely crucial to distinguish accident from design. We demand answers to such questions as, Did she fall, or was she pushed? Did someone die accidentally or commit suicide? Was this song conceived independently, or was it plagiarized? Did someone just get lucky on the stock market, or was there insider trading?

Not only do we demand answers to such questions, but entire industries are devoted to drawing the distinction between accident and design. Here we can include forensic science, intellectual property law, insurance claims investigation, cryptography, and random number generation—to name but a few. Science itself needs to draw this distinction to keep itself honest. Just last January there was a report in *Science* that a Medline Web search uncovered a "paper published in *Zentralblatt für Gynäkologie* in 1991 [containing] text that is almost identical to text from a paper published in 1979 in the *Journal of Maxillofacial Surgery.*"[3] Plagiarism and data falsification are far more common in science than we would like to admit. What keeps these abuses in check is our ability to detect them.

If design is so readily detectable outside science, and if its detectability is one of the key factors keeping scientists honest, why should design be barred from the content of science? Why do Dawkins and Crick feel compelled to remind us constantly that biology studies things that only appear to be designed, but that in fact are not designed? Why couldn't biology study things that are designed?

The biological community's response to these questions has

been to resist design absolutely. The worry is that for natural objects (unlike human artifacts) the distinction between design and nondesign cannot be reliably drawn. Consider, for instance, the following remark by Darwin in the concluding chapter of his *The Origin of Species:*

> Several eminent naturalists have of late published their belief that a multitude of reputed species in each genus are not real species; but that other species are real, that is, have been independently created...Nevertheless they do not pretend that they can define, or even conjecture, which are the created forms of life, and which are those produced by secondary laws. They admit variation as a *vera causa* in one case, they arbitrarily reject it in another, without assigning any distinction in the two cases.[4]

Biologists worry about attributing something to design (here identified with creation) only to have it overturned later; this widespread and legitimate concern has prevented them from using intelligent design as a valid scientific explanation.

Though perhaps it was justified in the past, this worry is no longer tenable. There now exists a rigorous criterion—complexity-specification—for distinguishing intelligently caused objects from unintelligently caused ones. Many special sciences already use this criterion, though in a pretheoretic form (forensic science, artificial intelligence, cryptography, archeology, and the Search for Extra-Terrestrial Intelligence). The great breakthrough in philosophy of science and probability theory of recent years has been to isolate and make precise this criterion. Michael Behe's criterion of irreducible complexity for establishing the design of biochemical systems is a special case of the complexity-specification criterion for detecting design.

What does this criterion look like? Although a detailed explanation and justification are fairly technical, the basic idea is straightforward and easily illustrated.[5] Consider how the radio astronomers in the movie *Contact* detected an extraterrestrial intelligence. This movie, which was based on a novel by Carl Sagan, was an enjoyable piece of propaganda for the SETI research program—the Search for Extra-Terrestrial Intelligence. In the movie, the SETI researchers found extraterrestrial intelligence. (The nonfictional researchers have not been so successful.)

How, then, did the SETI researchers in *Contact* find an extraterrestrial intelligence? SETI researchers monitor millions of radio signals from outer space. Many natural objects in space (such as pulsars) produce radio waves. Looking for signs of design among all these naturally produced radio signals is like looking for a needle in a haystack. To sift through the haystack, SETI researchers run the signals they monitor through computers programmed with pattern-matchers. As long as a signal doesn't match one of the preset patterns, it will pass through the pattern-matching sieve (even if it has an intelligent source). If, on the other hand, it does match one of these patterns, then, depending on the pattern matched, the SETI researchers may have cause for celebration.

The SETI researchers in *Contact* found the following signal:

1101110111110111111101111111111101111111111111011
1 1 1 1 1 1 1 1 1 1 1 1 1 1 1 0 1 1 1 1 1 1 1
1111111111110111111111111111111111110111111111111
1 1 1 1 1 1 1 1 1 1 1 1 1 1 1 1 1 0 1 1 1 1 1
1111111111111111111111111101111111111111111111111
1 1 1 1 1 1 1 1 1 1 1 1 1 1 1 0 1 1 1 1 1 1 1
1111111111111111111111111111111111011111111111111
1 1 1 1 1 1 1 1 1 1 1 1 1 1 1 1 1 1 1 1 1 1 1
1111110111111111111111111111111111111111111111111
1 1 1 1 1 0 1 1 1 1 1 1 1 1 1 1 1 1 1 1 1 1 1
1111111111111111111111111111111111110111111111111
1 1 1 1 1 1 1 1 1 1 1 1 1 1 1 1 1 1 1 1 1 1 1
1111111111111111111111111101111111111111111111111
1 1 1 1 1 1 1 1 1 1 1 1 1 1 1 1 1 1 1 1 1 1 1
1111111111111111111111011111111111111111111111111

1 1 1 1 1 1 1 1 1 1 1 1 1 1 1 1 1 1 1 1 1 1 1
1111111111111111111111011111111111111111111111111
1 1 1 1 1 1 1 1 1 1 1 1 1 1 1 1 1 1 1 1 1 1 1
1111111111111111111111110111111111111111111111111
1 1 1 1 1 1 1 1 1 1 1 1 1 1 1 1 1 1 1 1 1 1 1
1111111111111111111111111111111101111111111111111
1 1 1 1 1 1 1 1 1 1 1 1 1 1 1 1 1 1 1 1 1 1 1
1111111111111111111111111111111111111111111101111
1 1 1 1 1 1 1 1 1 1 1 1 1 1 1 1 1 1 1 1 1 1 1
1111111111111111111111111111111111111111111111111
1 1 1 1 1 1 1 1 1 1 1 1 1 0 1 1 1 1 1 1 1 1 1
1111111111111111111111111111111111111111111111111
1 1 1 1 1 1 1 1 1 1 1 1 1 1 1 1 1 1 1 1 1 1 1
1111111111111111011111111111111111111111111111111
1 1 1 1 1 1 1 1 1 1 1 1 1 1 1 1 1 1 1 1 1 1 1
111111111111111111111111111111111111111111111

In this sequence of 1126 bits, 1s correspond to beats and 0s to pauses. This sequence represents the prime numbers from 2 to 101, where a given prime number is represented by the corresponding number of beats (that is, 1s), and the individual prime numbers are separated by pauses (0s).

The SETI researchers in *Contact* took this signal as decisive confirmation of an extraterrestrial intelligence. What is it about this signal that decisively indicates design? Whenever we infer design, we must establish two things—complexity and specification. Complexity ensures that the object in question is not so simple that it can readily be explained by chance. Specification ensures that this object exhibits the type of pattern that is the trademark of intelligence.

To see why complexity is crucial for inferring design, consider the following sequence of bits:

110111011111

These are the first twelve bits in the previous sequence representing the prime numbers 2, 3, and 5 respectively. Now it is a sure bet that no SETI researcher, if confronted with this twelve-bit sequence, is going to contact the science editor at the *New York Times*, hold a press conference, and announce that an extraterrestrial intelligence has been discovered. No headline is going to read, "Aliens Master First Three Prime Numbers!"

The problem is that this sequence is much too short (has too little complexity) to establish that an extraterrestrial intelligence with knowledge of prime numbers produced it. A randomly beating radio source might by chance just happen to put out the sequence "110111011111." A sequence of 1126 bits representing the prime numbers from 2 to 101, however, is a different story. Here the sequence is sufficiently long (has enough complexity) to confirm that an extraterrestrial intelligence could have produced it.

Even so, complexity by itself isn't enough to eliminate chance and indicate design. If I flip a coin one thousand times, I'll participate in a highly complex (or what amounts to the same thing, highly improbable) event. Indeed, the sequence I end up flipping will be one in a trillion trillion trillion...where the ellipsis needs twenty-two more "trillions." This sequence of coin tosses won't, however, trigger a design inference. Though complex, this sequence won't exhibit a suitable pattern. Contrast this with the sequence representing the prime numbers from 2 to 101. Not only is this sequence complex, it also embodies a suitable pattern. The SETI researcher who in the movie *Contact* discovered this sequence put it this way: "This isn't noise; this has structure."

What is a *suitable* pattern for inferring design? Not just any pattern will do. Some patterns can legitimately be employed to infer design whereas others cannot. It is easy to see the basic intuition here. Suppose an archer stands fifty meters from a large wall with bow and arrow in hand. The wall, let's say, is sufficiently large that the archer can't help but hit it. Now suppose each time the archer shoots an arrow at the wall, the archer paints a target around the arrow so that the arrow sits squarely in the bull's-eye. What can be concluded from this scenario? Absolutely nothing about the archer's ability as an archer. Yes, a pattern is being matched, but it is a pattern fixed only after the arrow has been shot. The pattern is thus purely ad hoc.

But suppose instead the archer paints a fixed target on the wall and then shoots at it. Suppose the archer shoots a hundred arrows, and each time hits a perfect bull's-eye. What can be concluded from this second scenario? Confronted with this second scenario we are obligated to infer that here is a world-class archer, one whose shots cannot legitimately be explained by luck, but rather must be explained by the archer's skill and mastery. Skill and mastery are, of course, instances of design.

Like the archer who fixes the target first and then shoots at it, statisticians set what is known as a *rejection region* prior to an experiment. If the outcome of an experiment falls within a rejection region, the statistician rejects the hypothesis that the outcome is due to chance. The pattern doesn't need to be given prior to an event to imply design. Consider the following cipher text: *nfuijolt ju jt mjlf b xfbtfm.*

Initially this looks like a random sequence of letters and spaces—initially you lack any pattern for rejecting chance and inferring design.

But suppose next that someone comes along and tells you to treat this sequence as a Caesar cipher, moving each letter one notch back in the alphabet. Behold, the sequence now reads: *methinks it is like a weasel.*

Even though the pattern is now given after the fact, it still is the right sort of pattern for eliminating chance and inferring design. In contrast to statistics, which always tries to identify its patterns before an experiment is performed, cryptanalysis must discover its patterns after the fact. In both instances, however, the patterns are suitable for inferring design.

Patterns divide into two types, those that in the presence of complexity warrant a design inference and those that despite the presence of complexity do not warrant a design inference. The first type of pattern is called a *specification*, the second a *fabrication.*

Specifications are the non-ad hoc patterns that can legitimately be used to eliminate chance and warrant a design inference. In contrast, fabrications are the ad hoc patterns that cannot legitimately be used to warrant a design inference. This distinction between specifications and fabrications can be made with full statistical rigor.[6]

Why does the complexity-specification criterion reliably detect design? To answer this, we need to understand what it is about intelligent agents that makes them detectable in the first place. The principal characteristic of intelligent agency is choice. Whenever an intelligent agent acts, it chooses from a range of competing possibilities.

This is true not just of humans and extraterrestrial intelligences, but of animals as well. A rat navigating a maze must choose whether to go right or left at various points in the maze. When SETI researchers attempt to discover intelligence in the radio transmissions they are monitoring, they assume an extraterrestrial intelligence could have chosen to transmit any number of possible patterns, and then they attempt to match the transmissions they observe with the patterns they seek. Whenever a human being utters meaningful speech, he chooses from a range of utterable sound-combinations. Intelligent agency always entails discrimination—choosing certain things, ruling out others.

Given this characterization of intelligent agency, how do we recognize that an intelligent agent has made a choice? A bottle of ink spills accidentally onto a sheet of paper; someone takes a fountain pen and writes a message on a sheet of paper. In both instances ink is applied to paper. In both instances one among an almost infinite set of possibilities is realized. In both instances one contingency is actualized and others are ruled out. Yet in one instance we ascribe agency; in the other, chance.

What is the relevant difference? Not only do we need to observe that a contingency was actualized, but we ourselves need also to be able to specify that contingency. The contingency must conform to an independently given pattern, and we must be able independently to formulate that pattern. A random inkblot is unspecifiable; a message written with ink on paper is specifiable. Wittgenstein in *Culture and Value* made the same point: "We tend to take the speech of a Chinese for inarticulate gurgling. Someone who understands Chinese will recognize language in what he hears."[7]

In hearing a Chinese utterance, someone who understands Chinese not only recognizes that one from a range of all possible utterances was actualized, but he is also able to identify the utterance as coherent Chinese speech. Contrast this with someone who does not understand Chinese. He will also recognize that one from a range of possible utterances was actualized, but this time, because he lacks the ability to understand Chinese, he is unable to tell whether the utterance was coherent speech.

To someone who does not understand Chinese, the utterance will appear gibberish. *Gibberish*—the utterance of nonsense syllables uninterpretable within any natural language—always actualizes one utterance from the range of possible utterances. Nevertheless, gibberish, by corresponding to nothing we can understand in any language, also cannot be specified. As a result, gibberish is never taken for intelligent communication, but always for what Wittgenstein calls "inarticulate gurgling."

Experimental psychologists who study animal learning and behavior employ a similar method. To learn a task an animal must acquire the ability to actualize behaviors suitable for the task as well as the ability to rule out behaviors unsuitable for the task. Moreover, for a psychologist to recognize that an animal has learned a task, it is necessary not only to observe the animal making the appropriate discrimination, but also to specify this discrimination.

Thus to recognize whether a rat has successfully learned how to traverse a maze, a psychologist must first specify which sequence of right and left turns conducts the rat out of the maze. No doubt, a rat randomly wandering a maze also discriminates a sequence of right and left turns. But by randomly wandering the maze, the rat gives no indication that it can discriminate the appropriate sequence of right and left turns for exiting the maze. Consequently, the psychologist studying the rat will have no reason to think the rat has learned how to traverse the maze. Only if the rat executes the sequence of right and left turns specified by the psychologist will the psychologist recognize that the rat has learned how to traverse the maze.

Note that complexity is implicit here as well. To see this, consider again a rat traversing a maze, but now take a very simple maze in which two right turns conduct the rat out of the maze. How will a psychologist studying the rat determine whether it has learned to exit the maze? Just putting the rat in the maze will not be enough. Because the maze is so simple, the rat could by chance just happen to take two right turns, and thereby exit the maze. The psychologist will therefore be uncertain whether the rat actually learned to exit this maze or whether the rat just got lucky.

But contrast this now with a complicated maze in which a rat must take just the right sequence of left and right turns to exit the maze. Suppose the rat must take one hundred appropriate right and left turns, and that any mistake will prevent the rat from exiting the maze. A psychologist who sees the rat take no erroneous turns and in short order exit the maze will be convinced that the rat has indeed learned how to exit the maze, and that this was not dumb luck.

This general scheme for recognizing intelligent agency is but a thinly disguised form of the complexity-specification criterion. In general, to recognize intelligent agency we must observe a choice among competing possibilities, note which possibilities were not chosen, and then be able to specify the possibility that was chosen. What's more, the competing possibilities that were ruled out must be live possibilities and sufficiently numerous (hence complex) so that specifying the possibility that was chosen cannot be attributed to chance.

All the elements in this general scheme for recognizing intelligent agency (that is, choosing, ruling out, and specifying) find their counterpart in the complexity-specification criterion. It follows that this criterion formalizes what we have been doing right all along when we recognize intelligent agency. The complexity-specification criterion pinpoints what we need to be looking for when we detect design.

Perhaps the most compelling evidence for design in biology comes from biochemistry. In a recent issue of *Cell*, Bruce Alberts, president of the National Academy of Sciences, remarked:

> The entire cell can be viewed as a factory that contains an elaborate network of interlocking assembly lines, each of which is composed of large protein machines...Why do we call the large protein assemblies that underlie cell function *machines?* Precisely because, like the machines invented by humans to deal efficiently with the macroscopic world, these protein assemblies contain highly coordinated moving parts.[8]

Even so, Alberts sides with the majority of biologists in regarding the cell's marvelous complexity as only apparently designed. The Lehigh University biochemist Michael Behe disagrees. In *Darwin's Black Box*, Behe presents a powerful argument for actual design in the cell. Central to his argument is his notion of irreducible complexity. A system is irreducibly complex if it consists of several interrelated parts so that removing even one part completely destroys the system's function. As an example of irreducible complexity, Behe offers the standard mousetrap. A mousetrap consists of a platform, a hammer, a spring, a catch, and a holding bar. Remove any one of these five components, and it is impossible to construct a functional mousetrap.

Irreducible complexity needs to be contrasted with *cumulative complexity*. A system is cumulatively complex if the components of the system can be arranged sequentially so that the successive removal of components never leads to the complete loss of function. An example of a cumulatively complex system is a city. It is possible successively to remove people and services from a city until one is down to a tiny village—all without losing the sense of community, the city's "function."

From this characterization of cumulative complexity, it is clear that the Darwinian mechanism of natural selection and random mutation can readily account for cumulative complexity. Darwin's account of how organisms gradually become more complex as favorable adaptations accumulate is the flip side of the city in our example from which people and services are removed. In both cases, the simpler and more complex versions both work, only less or more effectively.

But can the Darwinian mechanism account for irreducible complexity? Certainly; if selection acts with reference to a goal, it can produce irreducible complexity. Take Behe's mousetrap. Given the goal of constructing a mousetrap, one can specify a goal-directed selection process that in turn selects a platform, a hammer, a spring, a catch, and a holding bar, and at the end puts all these components together to form a functional mousetrap. Given a prespecified goal, selection has no difficulty producing irreducibly complex systems.

But the selection operating in biology is Darwinian natural selection. And by definition this form of selection operates without goals, has neither plan nor purpose, and is wholly undirected. The great appeal of Darwin's selection mechanism was, after all, that it would eliminate teleology from biology. Yet by making selection an undirected process, Darwin drastically reduced the type of complexity biological systems could manifest. Henceforth biological systems could manifest only cumulative complexity, not irreducible complexity.

As Behe explains in *Darwin's Black Box:*

> An irreducibly complex system cannot be produced...by slight, successive modifications of a precursor system, because any precursor to an irreducibly complex system that is missing a part is by definition nonfunctional...Since natural selection can only choose systems that are already working, then if a biological system cannot be produced

> gradually it would have to arise as an integrated unit, in one fell swoop, for natural selection to have anything to act on.[10]

For an irreducibly complex system, function is attained only when all components of the system are in place simultaneously. It follows that natural selection, if it is going to produce an irreducibly complex system, has to produce it all at once or not at all. This would not be a problem if the systems in question were simple. But they're not. The irreducibly complex biochemical systems Behe considers are protein machines consisting of numerous distinct proteins, each indispensable for function. Together they are beyond what natural selection can muster in a single generation.

One such irreducibly complex biochemical system that Behe considers is the bacterial flagellum. The flagellum is a whiplike rotary motor that enables a bacterium to navigate through its environment. The flagellum includes an acid-powered rotary engine, a stator, O-rings, bushings, and a drive shaft. The intricate machinery of this molecular motor requires approximately fifty proteins. Yet the absence of any one of these proteins results in the complete loss of motor function.

The irreducible complexity of such biochemical systems cannot be explained by the Darwinian mechanism, nor indeed by any naturalistic evolutionary mechanism proposed to date. Moreover, because irreducible complexity occurs at the biochemical level, there is no more fundamental level of biological analysis to which the irreducible complexity of biochemical systems can be referred and at which a Darwinian analysis in terms of selection and mutation can still hope for success. Undergirding biochemistry is ordinary chemistry and physics, neither of which can account for biological information. Also, whether a biochemical system is irreducibly complex is a fully empirical question—individually knock out each protein constituting a biochemical system to determine whether function is lost. If so, we are dealing with an irreducibly complex system. Experiments of this sort are routine in biology.

The connection between Behe's notion of irreducible complexity and my complexity-specification criterion is now straightforward. The irreducibly complex systems Behe considers require numerous components specifically adapted to each other and each necessary for function. That means they are complex in the sense required by the complexity-specification criterion.

Specification in biology always makes reference in some way to an organism's function. An organism is a functional system comprising many functional subsystems. The functionality of organisms can be specified in any number of ways. Arno Wouters does so in terms of the *viability* of whole organisms, Michael Behe in terms of the *minimal function* of biochemical systems. Even Richard Dawkins will admit that life is specified functionally, for him in terms of the *reproduction* of genes. Thus in *The Blind Watchmaker* Dawkins writes, "Complicated things have some quality, specifiable in advance, that is highly unlikely to have been acquired by random chance alone. In the case of living things, the quality that is specified in advance is...the ability to propagate genes in reproduction."[11]

So there exists a reliable criterion for detecting design strictly from observational features of the world. This criterion belongs to probability and complexity theory, not to metaphysics and theology. And although it cannot achieve logical demonstration, it does achieve a statistical justification so compelling as to demand assent. This criterion is relevant to biology. When applied to the complex, information-rich structures of biology, it detects design. In particular, we can say with the weight of science behind us that the complexity-specification criterion shows Michael Behe's irreducibly complex biochemical systems to be designed.

What are we to make of these developments? Many scientists remain unconvinced. Even if we have a reliable criterion for detecting design, and even if that criterion tells us that biological systems are designed, it seems that determining a biological system to be designed is akin to shrugging our shoulders and saying that

God did it. The fear is that admitting design as an explanation will stifle scientific inquiry, that scientists will stop investigating difficult problems because they have a sufficient explanation already.

But design is not a science stopper. Indeed, design can foster inquiry where traditional evolutionary approaches obstruct it. Consider the term "junk DNA." Implicit in this term is the view that because the genome of an organism has been cobbled together through a long, undirected evolutionary process, the genome is a patchwork of which only limited portions are essential to the organism. Thus on an evolutionary view we expect a lot of useless DNA. If, on the other hand, organisms are designed, we expect DNA, as much as possible, to exhibit function. And indeed, the most recent findings suggest that designating DNA as "junk" merely cloaks our current lack of knowledge about function. For instance, in a recent issue of the *Journal of Theoretical Biology*, John Bodnar describes how "non-coding DNA in eukaryotic genomes encodes a language which programs organismal growth and development."[12] Design encourages scientists to look for function where evolution discourages it.

Or consider vestigial organs that later are found to have a function after all. Evolutionary biology texts often cite the human coccyx as a "vestigial structure" that hearkens back to vertebrate ancestors with tails. Yet if one looks at a recent edition of *Gray's Anatomy*, one finds that the coccyx is a crucial point of contact with muscles that attach to the pelvic floor. The phrase "vestigial structure" often merely cloaks our current lack of knowledge about function. The human appendix, formerly thought to be vestigial, is now known to be a functioning component of the immune system.

Admitting design into science can only enrich the scientific enterprise. All the tried and true tools of science will remain intact. But design adds a new tool to the scientist's explanatory tool chest. Moreover, design raises a whole new set of research questions. Once we know that something is designed, we will want to know how it was produced, to what extent the design is optimal, and what is its purpose. Note that we can detect design without knowing what something was designed for. There is a room at the Smithsonian filled with objects that are obviously designed but whose specific purpose anthropologists do not understand.

Design also implies constraints. An object that is designed functions within certain constraints. Transgress those constraints, and the object functions poorly or breaks. Moreover, we can discover those constraints empirically by seeing what does and doesn't work. This simple insight has tremendous implications not just for science but also for ethics. If humans are in fact designed, then we can expect psychosocial constraints to be hardwired into us. Transgress those constraints, and we, as well as our society, will suffer. There is plenty of empirical evidence to suggest that many of the attitudes and behaviors our society promotes undermine human flourishing. Design promises to reinvigorate that ethical stream running from Aristotle through Aquinas known as natural law.

By admitting design into science, we do much more than simply critique scientific reductionism. Scientific reductionism holds that everything is reducible to scientific categories. Scientific reductionism is self-refuting and easily seen to be self-refuting. The existence of the world, the laws by which the world operates, the intelligibility of the world, and the unreasonable effectiveness of mathematics for comprehending the world are just a few of the questions that science raises, but that science is incapable of answering.

Simply critiquing scientific reductionism, however, is not enough. Critiquing reductionism does nothing to change science. And it is science that must change. By eschewing design, science has far too long operated with an inadequate set of conceptual categories. This has led to a constricted vision of reality, skewing how science understands not just the world, but also human beings.

Martin Heidegger remarked in *Being and Time* that "a science's level of development is determined by the extent to which it is *capable* of a crisis in its basic concepts."[13] The basic concepts with which science has operated these last several hundred years are no longer adequate, certainly not in an information age, certainly not in an age where design is empirically detectable. Science faces a crisis of basic concepts. The way out of this crisis is to expand science to include design. To admit design into science is to liberate science, freeing it from restrictions that can no longer be justified.*

*"Science and Design" was first published in *First Things* 86 (October 1998): 21–27. It is reprinted here by permission of the author, Dr. William Dembski.

William Dembski

William Dembski, who has two earned doctorates in mathematics and the philosophy of science, is one of the leading theoreticians of Intelligent Design. He is especially known for his development of the "explanatory filter," a rigorous system of analysis that researchers are beginning to use in order to detect the action of an intelligent agent (or agents) in any object or phenomenon in the universe.

Dr. Dembski is research professor of mathematics at Baylor University in Waco, Texas. He is the author of dozens of articles and three books. *The Design Inference*, published by Cambridge University Press, set forth the mathematical basis for his explanatory filter. His ideas were further developed in *Intelligent Design*, from InterVarsity Press, and *No Free Lunch*, published by Rowman and Littlefield.

This chapter indicates that explanatory options in science must now be expanded. No longer can scientists be arbitrarily restricted to just two factors: "law" and "chance" (or contingency). A third explanatory option, "design," is now clearly needed to explain what we observe in nature. Professor Dembski explains why this is so.

Within the breast of nature throbs the heart of God.

—Anonymous

Conclusion

By James P. Gills, M.D.

Complexity, irreducible complexity, magnificent complexity. The cell and its DNA, nature's unbounded variation and variety, the eye, the pancreas, consciousness and imagination, and *spirit*—the mechanism of random emergence of traits selected by environment cannot account for the biology, the chemistry, the wonder of it all. Darwin could not—and Darwin's intellectual descendents still cannot—explain the emergence of phenomenon of life. Only an intelligence would have the adequate, appropriate palette and paints to generate such a masterpiece. The only explanation for both elm and elk is in design, an intelligent design, a design shrouded in mysterious beauty and predicated on a purpose.

Through the unique approaches of various chapters, our debate has focused attention on some principles that are either summarily dismissed or easily overlooked in a world predominated by unexamined evolutionary perspectives.

Darwinism and the Physical Evidence

Darwinism, in the face of physical evidence that we now possess, is untenable. We have seen that *microevolution*, the minute alterations that allow one individual to be born more capable than its siblings or enable a bacterium to survive a pharmacological assault, can be readily observed in the field. However, convincing evidence of *macroevolution*, the emergence of entirely novel creatures and of life itself through a mechanism of random mutation, is absent in the fossil evidence, embryology, and microbiology. And more, such evidence has not been forthcoming since Darwin first introduced his theory. Mathematical probabilities categorically rule out chance as the mediating factor in the appearance and diversification of life. Darwinism, therefore, can now be characterized as a "myth" that science has been unable to buttress sufficiently to win the wager on the theory's explanatory and predictive powers.

The only way to account for what we observe in nature is to assign the presence of a design and designer. In numerous examples, notably Behe's exposition of the bacterial flagellum and Thaxton's reflections on DNA, we have witnessed the thoroughgoing denuding of this Darwinian myth and, what's more, have seen the gleam of the nuts and bolts of a design there. The specified complexity in our genetic sequence, the requisite interdependence of constituent elements in a flagellum, integrated systems of tissue repair or vision, and the overwhelming complexity in a single cell all necessarily implicate an intelligent influence and inspiration in design. Yet, who is paying attention?

The imbalance that has held Darwinian theory as truth and the appeal to design as unfounded superstition is rooted in both an ignorance of recent, relevant scientific revelations and the spiritual shortsightedness of a species that refuses to look at God's own revelations of truth. Before science managed to peer into the black box of the cell, the appearance of design in nature was already evident to observers of all stripes, but it was rationalized by Darwin as merely *apparent* rather than actual design. Presently, however, the inclinations of those who believe in the design hypothesis are justified by an array of empirical evidences. However, one is still left to wonder how Darwinian theory has been capable of such tenacity in the ethos of so many reasonable people.

It has been said that the first point of wisdom is to know truth while the second is to discern what is false. Science in *its* search for truth has sought independence from dogma, ostensibly religious doctrine, and yet scientists have clung pertinaciously to another doctrine that is so silken and seductive, so integral and *unquestioned* as to be as much the organizing principle of an entire discipline as an explanatory device of individual instances. Many have failed to notice this intellectual contradiction. Perhaps, and this is becoming increasingly evident, they have actively *avoided* the contradiction in their espousal of evolution and natural selection.

The Emperor's New Clothes

> A little learning is a dangerous thing;
> Drink deep, or taste not the Pierian spring:
> There shallow draughts intoxicate the brain,
> And drinking largely sobers us again.
>
> —ALEXANDER POPE

Primarily, we can credit Darwinism with its capacity to hold an audience by the attraction of its internal logic and, perhaps more convincingly, as a countervailing philosophy to the possibility of divinity. How else are we to explain neo-Darwinist concessions that observed complexity in nature is highly improbable as a simple contingency, while reiterating that science will eventually find an explanation to its recalcitrant ways? Can this be interpreted as anything but a hesitance to open the eyes and face the deeper truth that obviously shines through the breaks and faults of a thin and fractured surface?

It is vital to note that shared, naturalistic assumption does not equal truth, especially when the assumption remains insulated from piercing inquiry. The nature of propaganda is to blind one eye to actual truth, and the nature of humanity is to hide from truth with self-inflicted blindness in the other eye. That potent admixture has long left us bumping into ideas in the dark though, it is clear, the light was always there. Public acquiescence to a group's shared blindness, fear of being the one to point out the emperor's nakedness, is as cogent an explanation of Darwin's long-standing dominance as any presumed scientific merit. Darwinism has too long been insulated by our prideful, self-inflicted blindness and our unwillingness to relinquish mastery to God.

The bias, dissonance really, in modern thinkers favoring Darwin is not exclusive to our present discussion of evolution vs. creation through intelligent design. Our debate merely exemplifies the main currents of intellectual tradition in the West, which, for the duration of several centuries, has affected the singular lives of individuals and the societies they are settled in, while dictating which beliefs merit legitimacy. The Word says that wisdom, and its attendant discernment, begin with fear of the Lord. (See Proverbs 1:7.) Humanity's science, like its religion, ought to be beholden to this canon, yet it has eschewed it in an intellectual ferment that severed faith and reason. That conceptual break has increasingly led to the intemperate debasement of faith and the reciprocal aggrandizement of reason, to the detriment of all.

As Michael Behe pragmatically states, we have to accept that an explanation of life within a model of intelligent design may be beyond the ability of science fully to explain. This is a proposition that sits uncomfortably in a frenetic human mind that cannot find its place of rest or ultimate wisdom within logical frameworks constructed on sensual indicators and interpretations. An important insight emerges: *We must be willing to acknowledge the limits of science.* The wisdom that accommodates a mind to its place in existence, that reveals the place of rest, is sourced in and leads back to God.

The Schism

The lure of evolutionary theory, taught formally as an unchallengeable scientific verity, countervails the attractive force of the gravity of Creation through God's design, as it is observed in the complexity of organisms. Within many people, this presents an irreconcilable tug-of-war or, perhaps more accurately, spiritual war. We sway, collectively and personally, pulled in disparate directions toward antithetical tendencies of mind and spirit. This marks our desire for the divine, engaged in conflict by and pitted against the tyranny of instrumental reason. This tyranny is both cause of and caused by a lack of humility. The release from the tension between these tendencies amounts to rest, the rest in God's Redemption—uncomplicated and liberating.

With the scientific revolution, intertwined with the Age of Reason, there was a break in Western intellectual tradition between a science increasingly viewed as enlightened and a religion taken to be anti-intellectual and reactionary. Prior to Galileo's cosmological "heresy," the medieval church had been the womb of scientific enquiry. However, precipitated by the Pyrrhic victory over Galileo and his opposing claims, a new philosophical belief formed. A political balance of power certainly shifted then, but more importantly, a whole philosophical worldview was irradicably altered. Subsequently, in the minds of laymen, as surely and tenaciously as within academic halls, science alone could provide a definitive, "true" perspective on reality. Anthropocentric orientations would come to be accepted as the sole foundation of truth.

Emerging from the animated blossoming of the Renaissance and Reformation, thinkers advanced the preeminence of reason in human experience. Voltaire evaluated the huge discrepancies between France's wealthy, sybaritic gentry and powerful clergy and the wretched masses. There he found fertile ground for theories expounding secular intellectualism and individualism that would influence all subsequent Western thought and unleash a dictatorship of reason. Descartes, the acknowledged father of modern philosophy, then liberated philosophy from theology with his *cogito* that grounded faith in God in subjective self-certainty. His views were instrumental in the nascent technological orientation of the modern world. David Hume, while claiming to be a theist, presented arguments in 1779 that undermined the classic proofs of God and, presaging Darwin, picked holes in the argument for design in nature. The influential Kant, in his desire to develop a view consistent with Newton's discoveries of natural laws, wrenched reason away from religious foundations altogether. Now rational speculation and empirical investigation were pictured as entirely independent of truth about God and corollaries of design hypotheses.

Into this intellectual flow waded Darwin. He was a convert to agnosticism, and his biology, accordingly, sought to reduce humanity to its constituent elements and to understand existence in the quantitative terms of sense data without reference to a greater calling or wisdom. Bear in mind that, as jarring as his theories were upon their release, this intellectual position was not unique. He was the quintessential voice of a Western intellectual, moral, and cultural climate that since the end of the Middle Ages had been progressively listing toward the mechanistic vision of both science and what it is to be human. The effects of this tilt are inescapably entrenched and massively far-reaching in our era. The vessel of civilization has been left without a keel.

It is in the personal rift between mind and spirit that many find the space, the loophole, that allows them to deny God's existence and impact in the universe and in their lives. Dazzled by the edifice of the scientific intellect, attracted to bright theoretical baubles, it is easy to take a path of less resistance and avoid the *responsibility* a creation has towards its Creator.

Civilization and Its Discontents

Philip Johnson's proposition in his "Darwinism on Trial" chapter clarifies a fine and easily missed distinction—the more important and essential issue is not the debate science has with creationism and intelligent design per se, but rather modern science's basis in materialist philosophy. The mystery of our biological origins is only an appendage to the overarching problem of life under the mandate of the Age of Reason and its fundamental shifts from religion and spirituality to science and the materialism of our senses. And where has that left civilization? Where has it left us?

Rationalism has bestowed the mantle of a secular priesthood upon scientists, technologists, and experts in our epoch. Promoting the supremacy of reason and in science's exclusive access to reliable knowledge of the world we inhabit has led to a most incredible ability to manipulate nature, to a life surrounded by the most astonishing technical marvels, on one hand. And on the other hand, this golden calf has deprived civilization of competing foundations of knowledge and wisdom and, thereby, of any method or attempt to control the path we are following. Who counsels us in our wanton manipulation of nature and destruction of environments, or our meddling with the genetic makeup of humans and other species?

Psychiatrist R. D. Laing poignantly described a young woman who, in her pain, lived in fear of the "bomb" she believed to be in her belly. He contrasted her with the "reasonable," "stable," and "well-adapted" men who coolly and dispassionately designed ever more destructive tools of war, the shredders of life and limb, as their vocation. The former was commonly considered insane, the latter sane, and Laing reasonably inquired how we reached the point that this could be possible. In our scientific age we rationalize absurdities and outrages while denying the Spirit that would guide judgment and advise prudence. With an unbalanced reliance upon our rationality, a relativistic utilitarianism has replaced more familiar ethical systems. The dignity of humanity has withered in the shadow of technology's *claim to authority* hidden from the clarity of God's light. As the masses tremble before their god of science and its liturgies of statistics, society slumps into alienation and anomie.

The legacy of Darwin's natural selection with its reductionist methodology has led to the development of sociobiology with its genetic explanations of our behavior. It has audaciously been imagined that society could be rationally planned, people rationally controlled. The Skinner box can indeed teach us something it did not intend—that within the confines of an increasingly rationalized and controlled environment the hapless inhabitants express themselves through the maximization of gain and advantage, through self-gratification and temporary release.

We see that the common good has largely dropped from view before numberless, myopic, and veiled eyes. The modern illusion of *apparent* individualism disguises what is actually the *conformism* of individuals—both as commodities (as ends in themselves) and consumers (economically and politically). Efficiency has reduced humans to material objects. Technology has lengthened our lives, enhanced our health, and given us greater access to *things* while simultaneously strangling the spiritual roots that would enable the full flowering of the individual and society. The price of amenity has been existential malaise.

Dissenting voices are muted by general, tacit agreement. The arcane machinery behind the pronouncements of science—some grandiose and some imaginary—is beyond the ken of most laymen. So we ponder our legitimacy to question and demur, given that we will be considered irrational and anti-intellectual. Here is the inevitable tug of war within and the struggle for preeminence by divergent views without. Loose relativism has left us bereft of higher moral laws and adrift in the social instability that is engendered there. As reason was afforded primacy within human experience, human reason has subsequently been endowed primacy within the universe. There is no appeal to, or even acknowledgment of, greater authority. Humanity, like some mad, minor, megalomanic dictator, has named itself and its science as lord, by its *own* authority. That is not even funny. Nor is it reasonable.

Providence has been recast in an inherently temporal, mortal, and *fallible* shape.

Sanctuary

What domain have we left God in the vastness of this universe? How do we answer for our pride that exalts human intellect and human achievement as apotheosis in Creation? And where is that pride propelling us? The Word warns us in Proverbs 16:18 that "pride goes before destruction, and a haughty spirit before a fall." Darwinism and scientific materialism have elevated humankind to a godlike status; their cold grip has denied us of any genuine aspirations that transcend the world that appears to our senses. The Truth asks that the uncritical embrace of naturalism, with its attendant rootlessness, relativism, loneliness, and sense of futility, be transformed into the abundant life that Jesus advocates. Rather than our own self-reliant, moral resolve, this transformation from self-centeredness to God-centeredness is driven by the love of the Designer and Savior, when we seek Him.

As we grapple with giant concepts of time, space, sub-atomic dimensions, and unimaginably complex cells, we are eternally challenged to find the invisible God in those places. Too often we are diverted by our senses so that "the cares of this world, the deceitfulness of riches, and the desires for other things entering in choke the word, and it becomes unfruitful" (Mark 4:19). If we deign to look, we will see God's eye in the sun, His gesture in the bee's waggle dance, and His arms in the ocean's fragrant embrace. The thumbprint of the Designer impressed in the DNA helix is likewise found in the inner man. The revealed beauty that evinces an immaculate design clothes our naked soul. Intimations of His presence are in all the bewildering details of the beauty, power, and precision in nature that our microscopes and telescopes can provide. But the true meaning of the design and the plan for each element in God's creation can only be apprehended by abandonment to Christ, by releasing faith in, and reliance upon, our own limited intelligence and the pride engendered by it.

The vision that opens a future to us is not a matter of empirical evidence, but faith in the news of God's redemption through Christ's sacrifice. Paul summed it up in Romans chapter 3:

> For all have sinned, and come short of the glory of God; being justified freely by his grace through the redemption that is in Jesus Christ.
>
> —ROMANS 3:23–24, KJV

And in this redemption we must trust and rest. God says, "In repentance and rest is your salvation, in quietness and trust is your strength" (Isa. 30:15, NIV). There we are adopted as the children of God. So, "Where then is the boasting? It is excluded" on the principle of faith. There is no place for our petty pride in God's glorious cosmos. Blindness begotten of pride in our own rational mind, reliance upon ourselves borne of fear in our existential insecurity, and the poverty of spirit that denies the sovereignty of God in His universe, all prevent us from finding the very sanctuary that we long for.

Augustine understood the limitations of human philosophy and human wisdom, and knew that "our hearts are restless till they rest in thee, O Lord." Yet, we do not willingly rest until we sense our need for God's redemption where we can experience His grace. We must become desperate for God. Abiding tragedy and calamity often may be the only way that we allow ourselves to seek the reality of God. Frequently when we turn to Him, we meet a barrier. Our sin and guilt alienate us from God's holiness, and our pride prevents us from seeking God's assistance through that barrier. In Luke 18, Jesus contrasted the humble tax collector, broken by a burden of contrition, who cried out, "God, be merciful to me a sinner," with the proud, sanctimonious Pharisee who thanked God because he was "not like other men" (vv. 13, 11). Jesus' response was, "For everyone who exalts himself will be humbled, and he who humbles himself will be exalted" (v. 14). The barrier between

God and man, between Creator and created, stands until we begin to rest on Christ's finished work.

God's might, majesty, and mastery are evident all about us. The irreducible complexity of an eye ought to incite genuine wonder and awe within us. Our refusal to acknowledge what God has to reveal there, in the cell, in our hearts, in the stars, and in His Word is our greatest mistake. Only in awe and acknowledgment can we authentically appreciate a gardenia-scented zephyr and the joy of being alive that it brings. The integration of living cells into a functioning body should remind us that we too are to be integrated into the body of Christ. As the cell lives in the Designer's plan, so a soul likewise abides. Darwinism and the materialism of humanity would make us each accidental, incidental, and meaningless. Through God's design, however, we become related, reliant, and replete with meaning. We are irreducibly complex, original, beautiful, and *integrated.* We are all one in the design and in our purpose to care for others and let the love of God flow through us to those around.

Are we willing to look past our noses and contemplate the pancreas? Are we willing to forgo the folly of prideful denial of both the divinity in our design and of our role and responsibility in Creation? When next a butterfly flutters by, will we look down or up? Science is necessarily an incomplete revelation of God's presence, but if we have the eyes to see, the truth is genuine, grand, and irrefutable. Moreover, in the gospel we have God's promise of love and redemption; we are offered forgiveness and acceptance into the fullness of His plan. There, free to embrace the comfort given by the Holy Comforter, we meet the Spirit of God as our inheritance and portion now and forever.

Appendix A

Science vs. Science

The debate over the teaching of evolution isn't just in Kansas anymore, as other states take up the issue. While these battles make headlines, they are the fruit of a scholarly movement that has shaken up the scientific establishment. WORLD talked to four Intelligent Design revolutionaries who are fighting Darwinists on their own terms.

By Lynn Vincent

The evolution debate reignited this month as Oklahoma Attorney General Drew Edmondson ruled that Oklahoma's State Textbook Committee doesn't have the authority to require that biology textbooks carry a disclaimer that calls Darwinism a "controversial theory." (Committee members plan to challenge the ruling.) Meanwhile, in Louisiana, the Tangipahoa School Board voted 5-4 against taking a defense of a similar disclaimer to the U.S. Supreme Court after an appeals court declared that the disclaimer is unconstitutional.

While none of this is good news for those who question Darwinism, one thing is clear: Darwinists are being forced to play defense. A major reason why is the emergence over the last few years of the Intelligent Design movement—a group of scholars and writers who argue that the world and its creatures show evidence of design. Who are some of the authors behind this movement? *WORLD* spoke with four of them.

Ignore That Man Behind the Curtain

In 1987, when UC Berkeley law professor Phillip Johnson asked God what he should do with the rest of his life, he didn't know he'd wind up playing Toto to the ersatz wizards of Darwinism. But a fateful trip by a London bookstore hooked Mr. Johnson on a comparative study of evolutionary theory. And by 1993, Mr. Johnson's book *Darwin on Trial* had begun peeling back the thin curtain of science that shielded evolution to reveal what lay behind: Darwinian philosophers churning out a powerful scientific mirage.

Darwin on Trial was the result of Mr. Johnson's years-long, lawyerly dissection of arguments for evolution. The forensic strategies of prominent evolutionists like Richard Dawkins and Stephen Jay Gould reminded Mr. Johnson of courtroom sleight-of-hand: Their materialist definition of terms decided the debate before opening arguments could begin. "I could see," he said, "that evolution was not so much science as a philosophy that Darwinists had adopted in the teeth of the facts."

Once evolutionists read his book, they were eager to sink their teeth into Mr. Johnson, whom they saw as a middle-aged, Harvard-educated dilettante sticking his unscientific nose where it didn't belong. Critics lined up to debate him. But once engaged, his adversaries found him to be both ruthlessly intelligent and maddeningly congenial. With his agreeable, favorite-uncle face, wire-rimmed specs, and a perpetual smile in his voice, it was hard not to like Mr. Johnson as he shredded their arguments. And, of all things, he even wanted to be friends when the debates were through.

"I've been overplayed as a controversialist," said Mr. Johnson, who sees such bridge building as his greatest strength. (God built a bridge to him during the failure of his first marriage, when he became a Christian believer. He met his second wife, Kathie, at a Presbyterian church conference.) "I see myself as a person who tries to build alliances and friendships. To win the debate, you have to carry both the moral high ground and the intellectual high ground rather than try to win by any sort of power tactics. That's really what we're trying to teach people."

The "we" is the cadre of Intelligent Design (ID) proponents for whom Mr. Johnson acted as an early fulcrum. In the early 1990s, as formidable scientists and theoreticians like Michael Behe, William Dembski, and others emerged in support of design theory, Mr. Johnson made contact, exchanged flurries of e-mail, and arranged personal meetings. He frames these alliances as a "wedge strategy," with himself as lead blocker and ID scientists carrying the ball in behind him.

"We're unifying the divided people and dividing the unified people," he said, adding that the "unified people" refers to Darwinists who at present occupy increasingly dissonant camps. The debate, he argues, is being successfully reformulated in a way that changes the balance of influence and "puts the right questions on the table."

Evidence of an influence shift comes in varied forms: For example, Paul Nelson, a graduate student in philosophy at the University of Chicago, was able to get approval for a Ph.D. dissertation arguing against the theory of common ancestry—a mighty feat at a liberal, secular university. And Baylor mathematician William Dembski is spearheading a conference in April at which heavy-hitting secular academics will present papers on both sides of the evolutionary argument.

Such double-edged debates delight Mr. Johnson. "The whole 'wedge' philosophy isn't that you present answers and people listen. It's that you get people debating the right questions, like 'How can you tell reason from rationalization?' and 'Can natural processes create genetic information?'" Mr. Johnson has just published a new book, *The Wedge of Truth*, a volume that frames fundamental questions he feels people ought to be debating in the controversy over origins.

"Once you get the right questions on the table," Mr. Johnson said, "you can relax a bit, because if people are discussing the right questions instead of the wrong ones, then the discussion will be moving in the direction of truth instead of away from it."

The Third Atom Bomb

The reeducation of Michael Behe began in a green recliner. On a chilly fall night in the same year Mr. Johnson was seeking direction from God, Mr. Behe, a professor of biochemistry at Pennsylvania's Lehigh University, sat at home in that recliner, transfixed by a book that shook the very foundations of his own understanding of science. It was three in the morning before he finished Michael Denton's book *Evolution: A Theory in Crisis* and turned out the lights. Nine years later, Mr. Behe himself published a book that began turning out the lights on the theory of evolution.

"Although I had pretty much believed evolution, because that's what I was taught, I always had an uneasy feeling and questions in my mind," said Mr. Behe, a Roman Catholic who grew up in a family of eight children in Harrisburg, Pennsylvania. "After reading Denton's book and seeing his rational, scientific approach to the problem, I decided I had signed on to something that just was not well supported. And, since evolution is such a strong component of many people's view of how the world works, I started to wonder: What else have I been told that is unsupported, or not true? It was a very intense, intellectual time."

That intensity ultimately gelled into *Darwin's Black Box* (Free Press, 1996), a book that hit secular scientists like an atom bomb. Charles Darwin himself had already provided a pass-fail test for his theory: "If it could be demonstrated that any complex organ existed which could not possibly have been formed by numerous, successive, slight modifications, my theory would absolutely break down." Mr. Behe's book (now in its sixteenth printing) was the first to administer Mr. Darwin's own test at the molecular level. Using simple yet scientifically bulletproof analyses, Mr. Behe showed that even at the cellular level many structures are "irreducibly complex," meaning that all parts of a structure have to be present in order for the structure to function at all. Thus, the slow, gradual changes proposed by Darwin were as likely to have led to the spontaneous formation of complex structures as are flour, sugar, eggs, and milk likely to gradually coalesce into a wedding cake.

Mr. Behe wrote: "Applying Darwin's test to the ultra-complex world of molecular machinery and systems that have been discovered over the past forty years, we can say that Darwin's theory has 'absolutely broken down.'"

Most of Mr. Behe's secular critics did not, of course, agree. His work has been the target of both scholarly rebuttal and brainless invective. But on the whole, *Darwin's Black Box* received surprisingly respectful treatment. Not only did many Christian groups name it one of the most important books of the twentieth century, but reporters from the mainstream press also flocked to Bethlehem, Pennsylvania, to see what made Mr. Behe tick. Secular universities slated him for speaking engagements. The venerable *New York Times* even shocked Mr. Behe by inviting him to submit an article explaining the main thesis of his book.

Still, Mr. Behe, who seems somewhat embarrassed that his name appears on "important author" lists with the likes of Tolkien and Solzhenitsyn, doesn't see himself as a scientific crusader. He doesn't look like one either. At a recent conference on intelligent design, the bearded Mr. Behe emerged as the Anti-Suit. Opting to take the podium in his usual uniform of a plaid shirt, blue jeans, and workboots, he looked, while lecturing, like what he is: a dad.

"I do not see myself as called to overturn thinking on evolution in the world," Mr. Behe said. "My primary focus is my marriage and my family. I see myself as called to raise my eight children, and anything else is gravy."

But what about having written a book that decimated the fallacious underpinnings of modern science? That, he allows with a smile, is pretty good gravy indeed.

God's Mathematician

It's easy to imagine what William Dembski's wife finds in the dryer lint trap after washing her husband's pants: equations. Long, elegant equations replete with tangents, vectors, and permutations tangled unceremoniously with tissue shreds in the lint trap. When Mr. Dembski speaks, equations come out. When he writes, equations come out. Surely he must keep a few spare equations in his pockets.

A mathematician with two Ph.D.s and director of Baylor University's Polanyi Center, an information theory research group, Mr. Dembski is a long string-bean of a man who would rather listen than speak. But swirling behind his glasses and thin, angular face is an intellect that helped vault intelligent design theory from the realm of the possible to the province of the probable. His book, *The Design Inference: Eliminating Chance Through Small Probabilities* (Cambridge University Press 1998), set secular scientists' skirts afire by crafting for the first time a scientifically rigorous "explanatory filter" for detecting design.

"In the scientific community, there is always the worry that when we make an attribution of design, that natural causes will end up explaining it," said Mr. Dembski, who is also a Discovery Institute senior fellow and the man whom author George Gilder once called "God's mathematician." "There's the sense that we 'can't do science' with design because we can't get a handle on it, or do it reliably. My work is aimed at refuting that view and showing that we can have a reliable criterion for detecting design and distinguishing it from other modes of explanations" of origins.

Mr. Dembski describes his own formative concept of origins as a "vague, theistic belief." The son of a biologist (he now lives in Waco, Texas, with wife, Jana, and two-year old daughter Chloe), he said, "There was a time when I accepted some form of evolutionary theory." But his understanding of God as the designer solidified early in his twenty-year Christian walk. Still, he points out that his theories—and intelligent design theory in general—spells designer with a small d. "Although I would personally identify God as the designer on theological grounds, the Bible is not entering into these discussions. Intelligent design theorists are trying to make it a fully rigorous, scientific enterprise."

As a result, Mr. Dembski sees not only a growing acceptance of ID theory among scientific faculty at Christian colleges, but also an emerging community of theistic academics at secular universities. But Massimo Pigliucci isn't one of them. A biologist, Mr. Pigliucci's sputtering, angry review of *The Design Inference* published in the journal *BioScience* called Mr. Dembski's work "trivial," "nonsensical," and "part of a large, well-planned movement whose object...is nothing less than the destruction of modern science."

Mr. Dembski loved it. "If the worst humiliation is not to be taken seriously, at least we're being taken seriously," adding that even fellow Darwinists panned Mr. Pigliucci's intemperate reaction to Mr. Dembski's book. "If we're generating such strong, visceral responses, we must be doing something right."

Making It Clear

When it comes to baby toys, Steve Meyer doesn't play favorites. Whether he's lecturing nineteen-year-old college freshmen or arguing for intelligent design before science elites, Mr. Meyer has no qualms about pressing together chains of brightly colored snap-lock beads or launching a superball across the room.

All, of course, in the name of science.

"I've found that most people, even scientists, don't mind having ideas made clear," said Mr. Meyer, a philosopher of science and a professor at Whitworth College in Spokane. "In intelligent design, making ideas clear is all to our advantage because the case for Darwinism really depends a lot on obfuscation. So, if [Darwinists] can conceal that with lots of difficult jargon and technical terminology, they can keep everybody but the experts out."

It's Mr. Meyer's aim to let the non-experts in. Tall, intense, and personable, he calls himself a "shameless popularizer" and is the acknowledged PR guy for the Design Movement. Speaking to a mixed group of scientists, philosophers, and journalists at a recent intelligent design conference in L.A., he blew up balloons and slapped magnetic letters on a child-sized whiteboard to simplify explanations of evidence for design in DNA. When he was through, the philosophers and journalists actually understood what he was talking about.

Mr. Meyer arrived at his own understanding of life's origins between shifts at Atlantic Richfield (ARCO) oilfields in Dallas. After graduating from Whitworth in 1980, Mr. Meyer went to work for ARCO as a geophysicist. In 1985, a conference convened in Dallas that brought together top philosophers, cosmologists, and biologists to discuss the interrelationship of recent scientific findings and religion. Mr. Meyer, who basically wandered in off the street to listen in, found his own vaguely held notions of theistic evolution dismantled by former big-gun Darwinists who had themselves concluded that scientific evidence pointed to an intelligent designer of the universe.

"For me, it was a seminal event, a turning point," Mr. Meyer said. "I saw that there was an exciting, intellectual program here worth pursuing." It was a turning point that would lead him to Cambridge University where, in 1991, he earned his doctorate in the history and philosophy of science for a dissertation on origin-of-life biology.

Now, Mr. Meyer divides his time between Whitworth and his position as director of the Seattle-based Center for the Renewal of Science and Culture. The center, says its mission statement, "seeks to challenge materialism on specifically scientific grounds." Mr. Meyer said the center was founded as an academic end-run around secular university research departments held hostage by Darwinists. With its corps of forty research fellows in disciplines ranging from genetics to biology to artificial intelligence, he contends the center has the academic firepower to engineer a profound shift in the naturalistic paradigm that now dominates the culture.

For his part, Mr. Meyer stays busy with fundraising, budget management, and his own research on the evidence for design in DNA. (His book, *DNA by Design*, will be published this year.) He also keeps design theory alive in public forums. For example, when last year's controversy regarding the teaching of evolution in Kansas erupted, Mr. Meyer debated evolutionary biologists on National Public Radio. And his science and op-ed pieces appear in major papers, including *The Wall Street Journal* and the *Los Angeles Times.*

Of course, his critics publish op-eds of their own. He, like his ID colleagues, is regularly slammed as "anti-scientific" and "anti-intellectual."

"The gatekeepers of evolutionary theory are very worried about the Design Movement," Mr. Meyer said. "It's got a huge appeal with students, it's framed in a way that makes the Darwinian position very unattractive, and the evidence supports it. When it was religion vs. science, evolutionists won that debate every time."

Now, it's *science vs. science*, he said. And what the debate evolutionists had thought was settled has only just begun.*

---

*This article is reprinted with permission, *WORLD Magazine*, copyright © 2002. For information, visit www.worldmag.com.

Appendix B

The Alabama Biology Textbook Disclaimer

The Alabama biology textbook disclaimer (conveyed to us by Judy Mansel of the Alabama Department of Education) reads:

> A Message From the Alabama State Board of Education:
>
> This textbook discusses evolution, a controversial theory some scientists present as a scientific explanation for the origin of living things, such as plants, animals, and humans.
>
> No one was present when life first appeared on earth. Therefore, any statement about life's origins should be considered as theory, not fact.
>
> The word *evolution* may refer to many types of change. Evolution describes changes that occur within a species. (White moths, for example, may "evolve" into gray moths.) This process is microevolution, which can be observed and described as fact. Evolution may also refer to the change of one living thing to another, such as reptiles into birds. This process, called macroevolution, has never been observed and should be considered a theory. Evolution also refers to the unproven belief that random, undirected forces produced a world of living things.
>
> There are many unanswered questions about the origin of life which are not mentioned in your textbook, including:

- Why did the major groups of animals suddenly appear in the fossil record (known as the "Cambrian Explosion")?
- Why have no new major groups of living things appeared in the fossil record for a long time?
- Why do major groups of plants and animals have no transitional forms in the fossil record?
- How did you and all living things come to possess such a complete and complex set of "instructions" for building a living body?

Study hard and keep an open mind. Some day you may contribute to the theory of how living things appeared on earth.

Notes

Acknowledgments

1. Richard A. Swenson, *More Than Meets the Eye: Fascinating Glimpses of God's Power and Design* (Colorado Springs, CO: NavPress, 2000).

Introduction

1. Michael J. Behe, *Darwin's Black Box: The Biochemical Challenge to Evolution* (New York: The Free Press, 1996).
2. The reader can refer to Thomas Woodward's chapter in this book, "Meeting Darwin's Wager," for a description of the events leading up to, and beyond, the publication of *Darwin's Black Box.* Behe's book was reviewed by the *New York Times* on August 4, 1996, and on October 29, the paper published Behe's first opinion piece with that paper, "Darwin Under the Microscope." (Behe also wrote an opinion piece in the *New York Times* in August 1999 during the Kansas furor.) Feature coverage was given in December 1997, after a team of design advocates appeared in a nationally televised debate on PBS. The most important coverage was a front-page article on Intelligent Design, written by Nicholas Wade, that appeared in April 9, 2001.
3. Dembski's work on the explanatory filter is found in many books and articles, and his chapter in this book (the next to last chapter in Part Two) is a good place to start. His most important presentations of the explanatory filter are in *Mere Creation* (InterVarsity Press), edited by Dembski himself, and in his three books published in recent years: *The Design Inference* (Cambridge University Press), *Intelligent Design* (InterVarsity Press), and *No Free Lunch* (Rowman and Littlefield).
4. The "shock heard round the world" in August 1999, when the Kansas state board of education made this decision, was effectively reversed early in 2001 when newly elected board members voted to return

Kansas to its previous policy—requiring the teaching of Darwinian macroevolution along with microevolution.

Chapter 3
The Magnificently Complex Cell

1. Behe, *Darwin's Black Box*, 31.
2. Ibid., x.
3. Swenson, *More Than Meets the Eye*, 17.
4. Ibid., 24.
5. Ibid., 29.
6. Ibid., 34.
7. Enzo Russo and David Cove, *Genetic Engineering: Dreams and Nightmares* (New York: W. H. Freeman, 1995).
8. Behe, *Darwin's Black Box*, 39.
9. W. Wayt Gibbs, "Cybernetic Cells," *Scientific American* (August 2001): 55.
10. Rupert Sheldrake, *A New Science of Life* (London: Blond and Briggs, 1981), 11.
11. Behe, *Darwin's Black Box*, 139.
12. Swenson, *More Than Meets the Eye*, 21.
13. John R. Cameron, James G. Skofronick, and Roderick M. Grant, *Physics of the Body* (Madison, WI: Medical Physics Publishing, 1999), 38. Quoted in Swenson, *More Than Meets the Eye*, 21.
14. David Rosevear, "The Myth of Chemical Evolution," *Impact* (July 1999): iv. Quoted in Swenson, *More Than Meets the Eye*, 21.
15. Mark Caldwell, "The Clock in the Cell," *Discover* (October 1998): 36. Quoted in Swenson, *More Than Meets the Eye*, 21.
16. Swenson, *More Than Meets the Eye*, 61.
17. Russo and Cove, *Genetic Engineering*, 33.
18. Source obtained from the Internet: "Primer on Molecular Genetics," Human Genome Project Information. Website: www.ornl.gov/hgmis/publicat/primer/prim1.html.
19. Ibid.
20. Swenson, *More Than Meets the Eye*, 65.
21. Ibid.

22. Ibid.
23. Gunjan Sinha, "Human Genome: Less Code, More Complexity," *Popular Science* (January 2002): 62.
24. Ibid.
25. Swenson, *More Than Meets the Eye*, 24.
26. Ibid.
27. Lewis Thomas, *The Lives of a Cell* (New York: Viking Press, 1974), 93.
28. Behe, *Darwin's Black Box*, 213.
29. Dean L. Overman, *A Case Against Accident and Self-Organization* (New York: Rowman and Littlefield, 1997), 187.
30. Behe, *Darwin's Black Box*, 168–169.
31. A. Paul Alvisatos, "Less Is More in Medicine," *Scientific American* (September 2001): 16.
32. Fred Hoyle, *The Intelligent Universe* (New York: Holt, Rinehart and Winston, 1983), 12.
33. Ibid.
34. Ibid., 17.
35. Overman, *A Case Against Accident and Self-Organization*, 59, emphasis added.
36. Harold J. Morowitz, *Energy Flow in Biology* (Woodbridge, CT: Ox Bow Press, 1979), Quoted in Overman, *A Case Against Accident and Self-Organization*, 63.
37. Chandra Wickramasinghe, "Threats on Life of Controversial Astronomer," *New Scientist* (January 21, 1982): 140. Quoted in Overman, *A Case Against Accident and Self-Organization*, 60.
38. Swenson, *More Than Meets the Eye*, 71; citing J. P. Moreland (Ed.) quoting Carl Sagan and Francis Crick in *The Creation Hypothesis: Scientific Evidence for an Intelligent Designer* (Downers Grove, IL: Intervarsity, 1994), 272.
39. A. Koestler, "Beyond Atomism and Holism" in *Beyond Reductionism*, Alpbach Symposium, 1968, A. Koestler and J. R. Smythies (eds.), (London: Hutchinson and Co., 1969), 195–196.
40. Hoyle, *The Intelligent Universe*, 47.
41. Ibid., 40.

42. Behe, *Darwin's Black Box*, 45.
43. Overman, *A Case Against Accident and Self-Organization*, 56.
44. Francis Crick, *Life Itself: Its Nature and Origin* (New York: Simon and Schuster, 1981), 88.
45. David L. Hull, *Philosophy of Biological Science* (Englewood Cliffs, NJ: Prentice Hall, 1974), 70.
46. Ibid., 49.

CHAPTER 4
DNA, DESIGN, AND THE ORIGIN OF LIFE

1. Williams Paley, *Natural Theology* (Boston MA: Gould, Kendall and Lincoln, [1802], 1835).
2. Charles Darwin, *The Origin of Species* (New York: Washington Square Press, [1859], 1963).
3. James S. Trefil, *The Moment of Creation*, (New York: Charles Scribner's Sons, 1983).
4. Sir Karl Popper and Sir John Eccles, *The Self and Its Brain* (New York: Springer-Verlag, 1977).
5. Claude E. Shannon and Warren Weaver, *The Mathematical Theory of Communication* (Urbana, IL: University of Illinois Press, 1964).
6. J. D. Watson and F. H. C. Crick, "The Structure of DNA," *Cold Spring Harbor Symposium Quantitative Biology* 18 (1953):123.
7. Hubert P. Yockey, "Self Organization Origin of Life Scenarios and Information Theory," *J. Theoret*, Biol. 91, (1981): 13, 16.
8. C. Thaxton, W. Bradley, and R. Olsen, *The Mystery of Life's Origin* (New York: Philosophical Library, 1984), see Epilogue.
9. See the opening chapter and the chapter on molecular biology in N. Pearcey and C. Thaxton, *The Soul of Science* (Wheaton, IL: Crossway Books, 1994).
10. *Origins* (The Bhaktivedanta Institute, 3764 Watseka Avenue, Los Angeles, CA 90034, 1984), 38. Copies may be purchased from a local Hare Krishna representative or by contacting the Institute directly. "Many people think that the only alternative to Darwinism evolution would be some form of Biblical creationism. There are, however, many alternatives, including concepts of a universal designing intel-

ligence other that the one advocated by fundamentalist Christians and concepts of evolution other than the one advocated by Darwin."

11. Cicero, *The Nature of the Gods.* Translated by Horace C. P. McGregor with an introduction by J. M. Ross (New York: Penguin Books, 1972). See especially Book II, which is devoted to expounding the Stoic view. The Stoics were pantheists.
12. Isaac Newton, "General Scholium," *Mathematical Principles of Natural Philosophy*, R.M. Hutchins, ed. (Chicago: Great Books of the Western World, [1687], 1952). 369.
13. Frederick Ferre, "Design Argument," *Dictionary of the History of Ideas*, Vol. I. (New York: Charles Scribner's Sons, 1973), 673.
14. The person who accepts an intelligent cause might respond that this merely pushes the question back a step. We still have to account for the order built into water molecules themselves, which causes them to crystallize in such delightful patterns. Where did this order come from?
15. Actually, the "chip marks" were not due to conventional chisels of artisans. Small charges of explosive were used to blast away the unwanted granite, merely adding to the wonder of this national monument.
16. The term comes from Leslie Orgel, *The Origins of Life* (New York: John Wiley & Sons, 1973), 189.
17. Ibid., 190.
18. Ibid., 189.
19. The mathematical basis for distinguishing order from complexity is given by H. Yockey, "A Calculation of the probability of Spontaneous Biogenesis by Information Theory," *I Theoret, Biol* 67 (1977): 377.
20. In Henry Quastler's colorful expression, it is an "accidental choice remembered." Henry Quastler, *The Emergence of Biological Organization* (New Haven, CT: Yale University Press, 1964), 15–17.
21. No different, that is, merely in terms of nucleotide base sequences. There are other differences, however, between the DNA of living things and random mixtures of nucleotides. The greatest difference is that the DNA of living things contains the optically active, right-handed, sugar d-deoxyribose whereas a random mixture contains equal amounts of both d- and I-, or right- and left-handed forms.

22. Yockey, *The Emergence of Biological Organization*, 380.
23. In the case of ancient Egyptian hieroglyphics, scholars could only guess that the markings contained a message until the Rosetta Stone was discovered as a decoding key. Likewise, without such a decoding key a message from space might never be recognized, and thus never understood.
24. "High" and "low" are qualitative designations. A more quantitative way is to note that for the universe of 1,080 fundamental particles there are about 1080 instructions (bits of information) required to specify it. Thus the inorganic universe (without life) seems to have a lot of information. However, it is "low" when compared to *E. coli* bacterium, which has an information store of about 102,000,000 bits. Living structures have so much more information than the inorganic universe because, unlike the inorganic universe, the sequence in which its components are assembled is essential to its function.
25. Opposed to this is the "*manana* argument"—tomorrow we will find a natural cause. However, since we already have clear, unmistakable evidence for intelligent production of specified complexity, the burden is on opponents to show that a natural cause can also produce it.
26. In addition to the extensive review of results in *The Mystery of Life's Origin* (see note 8), interested readers should consult Robert Shapiro, *Origins: A Skeptic's Guide to the Creation of Life on Earth* (New York: Summit Books, 1986).
27. Fred Hoyle and Chandra Wickramasinghe, *Evolution From Space*, (New York: Simon and Schuster, 1981). Note that Hoyle and Wjckramasinghe do not argue for an intelligent cause, but against a natural cause.
28. David Hume, "An Enquiry Concerning Human Understanding," *Great Books of the Western World*, R. M. Hutchins, ed. (Chicago, [1748], 1952), 462, 499. "From causes which appear similar we expect similar effects...the same rule holds, whether the cause assigned be brute unconscious matter, or a rational intelligent being."
29. Note the similarity to attempts to determine whether extraterrestrial contact with earth has taken place in the past. Says Carl Sagan: "...convincing would be a certain class of artifact. If an artifact of technology

were passed on from an ancient civilization—an artifact that is far beyond the technological capabilities of the originating civilization—we would have an interesting prima facie case for extraterrestrial visitation." Carl Sagan, *The Cosmic Connection* (Garden City, NY: Anchor Books, 1973), 205.

30. A fanciful story about the search for extraterrestrial intelligence (SETI), but accurate as to the procedure, is: Carl Sagan, *Contact* (New York: Simon and Schuster, 1985).
31. Hume's argument from "uniform experience" has found wide application. David Hume, the skeptic, used the argument against miracles. "There must, therefore, be a uniform experience against every miraculous event." *Enquiry*, 491.
32. Michael Polanyi, "Life Transcending Chemistry and Physics," *Chemistry & Engineering News* (August 21, 1967): 54.
33. M. Eigen, W. Gardiner, P. Schuster, and R. Winkler-Oswatitsch, "The Origin of Genetic Information," *244* 4 (1981): 88.
34. G. Nicolis and I. Prigogine, *Self-Organization in Nonequilibrium Systems* (New York: Wiley Interscience, 1977).
35. A. Babloyantz, *Molecules, Dynamics, and Life* (New York: John Wiley & Sons, Inc., 1986).
36. S. W. Fox and K. Dose, *Molecular Evolution and the Origin of Life* (San Francisco, CA: W.H. Freeman, 1972).

CHAPTER 5
DARWINISM ON TRIAL

1. In February 1990, Johnson met with likeminded skeptics of Darwinism at an informal weekend conference at a hotel in Portland, Oregon. At this point, the group was known by the whimsical name, the "Ad Hoc Origins Committee." This group, after six more years, grew and eventually accepted the name "Intelligent Design." Thus, Johnson led the young movement, even before it was known by that title.
2. The phrase in quotes is Johnson's own invention, one that he uses with some frequency in his public lectures to sum up his evaluation of the evidence for macroevolution.
3. See S. J. Gould's critique of the "artifact theory" of the Precambrian

fossil record—and its distorting effect in inducing the "Burgess shoehorn"—in *Wonderful Life: The Burgess Shale and the Nature of History* (New York: W. W. Norton & Company, 1990), 271–277.

4. See Dawkins's discussion of saltation in Richard Dawkins, *The Blind Watchmaker: Why the Evidence of Evolution Reveals a Universe Without Design* (New York: W. W. Norton & Company, 1986).
5. The logical jump described in this sentence is essentially a play upon the term "relationship." Human family relationships (siblings, cousins, etc.) are based on more or less recent common biological ancestry, and it is understandable but illogical to assume that the same must be true of the "relationship" between bats and whales, or mammals and birds. Common ancestry as an explanation for natural classification is entirely reasonable as a hypothesis, but when enshrined as an irrebuttable presumption it is a projection of commonsense prejudice. As the physicists have been telling us, materialist common sense is not necessarily a reliable guide to scientific truth.
6. George Gaylord Simpson, *The Meaning of Evolution: A Study of the History of Life and of Its Significance for Man* (New Haven, CT: Yale University Press, 1949).
7. This quote from Douglas Futuyma is not from a popular polemic, but is taken from Futuyma's college textbook.
8. Dawkins, *The Blind Watchmaker.*
9. William Provine, "Evolution and the Foundation of Ethics," *MBL Science*, Vol. 3, no. 1: 25–29.

Chapter 6
Of Canadian Oddballs and Chinese Monsters

1. J. Madeleine Nash, "When Life Exploded," *Time* cover story (December 4, 1995); "The Explosion of Life," *National Geographic* (October 1993).
2. Michael Denton, *Evolution: A Theory in Crisis* (Bethesda, MD: Adler & Adler, 1986). On pages 101–104 Denton quotes several fossil experts who wrote to Darwin, or critiqued his book publicly, in connection with the fossil evidence.
3. I have kept copies of Dr. Bonner's letters, and I enjoyed (along with

Charles Thaxton and Jon Buell) a wonderful hour together in person with Dr. Bonner in December 1988. His remark to me about "bursts" was made in a phone conversation in the spring of 1986.

4. Gould's theory (developed with Niles Eldredge and Steven Stanley) says that the sudden appearance of new types occurs in the fossil record because much of evolution occurred in sudden bursts, following long periods of stability (stasis). Gould's mechanism for this sudden change is principally one of physical isolation of a small breeding group (out at the periphery of a large main population) followed by the reception and rapid proliferation in that group of "macromutations" (sometimes called "systemic mutations") which may have served as a "key" to transform a whole part of the animals morphology in, say, 10,000 years (a blink of an eye in geological time) rather than millions of years. Once the change was complete, then, the small group managed to succeed, spread, and (ultimately) leave a fossilized sample of that rapid change. See Gould's book *The Panda's Thumb*, especially his essays "The Episodic Nature of Evolutionary Change" and "The Return of the Hopeful Monster."
5. S. J. Gould, "A Short Way to Big Ends," *Natural History Magazine* 95 (January 1986): 18–19.
6. I should add that scientists in China are now researching what seem to be fossilized sponges in layers below the Cambrian.
7. Darwin, *The Origin of Species*, 292. There are several other places in Darwin's book where he expressed the same concern about the puzzling absence of fossil transitions.

CHAPTER 7
MEETING DARWIN'S WAGER

1. Behe, *Darwin's Black Box*. To read Behe's favorite quote by Darwin, see page 15.
2. Scott Swanson, "Debunking Darwin?", *Christianity Today*, Vol. 41, No. 1 (January 6, 1997): 64.
3. David Berlinski, as recorded in a personal interview, where he reiterated and confirmed his viewpoint, expressed in the blurbs on the publisher's publicity sheet for the book.

4. John Tagliabue, "Pope Bolsters Church's Support for Scientific View of Evolution," *New York Times* (October 25, 1996): A1.
5. Michael J. Behe, "Darwin Under the Microscope," *New York Times* (October 29, 1996): ed/op, A25.
6. Jerry Coyne, "God in the Details," *Nature* (September 19, 1996): 227.
7. Denton, *Evolution: A Theory in Crisis*, 358.
8. "Briefings" News Section, *Science* (July 26, 1991).
9. "Letters to the Editor," *Science* (August 30, 1991).
10. Jon Buell and Virginia Hearns, eds., *Darwinism: Science or Philosophy?* (Dallas: Foundation for Thought and Ethics, 1994).

Chapter 8
The Modern Intelligent Design Hypothesis: Breaking Rules

1. John Maddox, "Down with the Big Bang," 2 340 (1989): 425.
2. Darwin, *The Origin of Species*, 154.
3. Behe, *Darwin's Black Box*, 239.
4. Christian DeDuve, *Vital Dust: Life as a Cosmic Imperative* (New York: Basic Books, 1995), xiv.
5. David J. DeRosier, "The Turn of the Screw: The Bacterial Flagellar Motor." *Cell* 93 (1998): 17–20.
6. Kenneth R. Miller, *Finding Darwin's God: A Scientist's Search for Common Ground between God and Evolution* (New York: Cliff Street Rooks, 1999), 146–47.
7. Ibid.
8. Barry G. Hall, "Experimental Evolution of Ebg Enzyme Provides Clues about the Evolution of Catalysis and to Evolutionary Potential," *FEMS Microbiology Letters* 174 (1999): 1–8.
9. "Evolution of a Regulated Operon in the Laboratory," *Genetics* 101 (1982): 335–344.
10. "Evolution on a Petri Dish: The Evolved ß-Galactosidase System as a Model for Studying Acquisitive Evolution in the Laboratory" in M.K. Hecht and G.T. Prance, eds., *Evolutionary Biology* (New York: Plenum Press, 1982), 85, 150.
11. See I. Cairns, "Mutation and Cancer: The Antecedents to Our

Studies of Adaptive Mutation," *Genetics* 148 (1998): 1433–1440; P.L. Foster "Mechanisms of Stationary Phase Mutation: A Decade of Adaptive Mutation," *Annual review of Genetics* 33 (1999): 57–55; Barry G. Hall, "Adaptive Mutagenesis: A Process That Generates Almost Exclusively Beneficial Mutations," *Genetica* 102/3 (1998): 109–125; J. McFadden and J. Al Khalili, "A Quantum Mechanical Model of Adaptive Mutation," *Biosystems* 50 (1999): 203–211; and J.A. Shapiro. "Genome Organization, Natural Genetic Engineering and Adaptive Mutation," *Trends in Genetics* 13 (1997): 98–104.

12. Barry G. Hall "On the Specificity of Adaptive Mutations," Genetics 145 (1997): 39–44.
13. See idem, "Evolution of a Regulated Operon."
14. See idem, "Experimental Evolution of Ebg Enzyme."
15. Ibid.
16. Idem, "Experimental Evolution of *Ebg* Enzyme," 1–8.
17. Russell F. Doolittle, "A Delicate Balance," *Boston Review* (Feb/March 1997): 28–29.
18. T. H. Bugge, K. W. Kombrinck, M. J. Flick, C. C. Daugherty, M. J. Danton, and J. L. Degen, "Loss of Fibrinogen Rescues Mice from the Pleiotropic Effects of Plasminogen Deficiency," *Cell* 87 (1996): 709–719.
19. Doolittle, "A Delicate Balance," 28–29
20. T. H. Bugge, Q. Xiao, K. W. Kombrinck, M. J. Flick, K. Holmback, M. J. Danton, M. C.Colbert, D. P. Witte, K. Fujikawa, E. W. Davic, J. L. Degen, "Fatal Embryonic Bleeding Events in Mice Lacking Tissue Factor, the Cell-Associated Initiator of Blood Coagulation," *Proceedings of the National Academy of Sciences of the United States of America* 93 (1996): 6258–63; W. Y. Sun, D. P. Witte, J. L. Degen, M. C. Colbert, M. C. Burkart. K. Holmback, Q. Xiao, T. H. Bugge, and S. J. Degen, "Prothrombin Deficiency Results in Embryonic and Neonatal Lethality in Mice," *Proceedings of the National Academy of Sciences of the United States of America* 95 (1998): 7597–7602.
21. National Academy of Sciences, *Science and Creationism: A View from The National Academy of Sciences* (Washington, DC: National Academy Press. 1999), 25.

22. Miller, *Finding Darwin's God*, 146-47.
23. Darwin, *The Origin of Species*, 154.

Chapter 9
On the Design of the Vertebrate Retina

1. W. Thwaites, "Design: Can we see the hand of evolution in the things it has wrought?" In *Evolutionists Confront Creationists*, Proceedings of the 63rd Annual Meeting of the Pacific Division, AAAS, Volume 1, part 3 (San Francisco: Pacific Division, AAAS), 1992, 206–213.
2. G. C. Williams, *Natural Selection: Domains, Levels, and Challenges* (Oxford: Oxford University Press, 1992).
3. J. Diamond, "Voyage of the Overloaded Ark," *Discover*, June 1985, 82–92.
4. Dawkins, *The Blind Watchmaker*.
5. R. F. Miller, "The physiology and morphology of the vertebrate retina," S. J. Ryan, ed., *Retina*, 2nd ed., Volume 1: Basic Science & Inherited Retinal Disease (St. Louis: Mosby, 1994), 58–71.
6. Williams, *Natural Selection*, 73.
7. Ibid., 72.
8. Diamond, "Voyage of the Overloaded Ark," 91.
9. Miller, "The physiology and morphology of the vertebrate retina," 30; see also Diamond, "Voyage of the Overloaded Ark," 91 and Williams, *Natural Selection*, 74.
10. K. D. Farnsworth and K. J. Niklas, "Theories of optimization, form and function in branching architecture of plants," *Functional Ecology* 9 (1995): 355–363.
11. T. H. Goldsmith, "Optimization, Constraint, and History in the Evolution of Eyes," *Quarterly Review of Biology* 65 (1990): 286.
12. See, for instance: R. H. Steinberg, "Interactions between the retinal pigment epithelium and the neural retina," *Documenta Opthalmologia* 60 (1985): 327–346; and K. M. Zinn and M. F. Marmor, *The Retinal Pigment Epithelium* (Cambridge, MA: Harvard University Press, 1979).
13. C. D. B. Bridges, "Distribution of retinal isomerase in vertebrate eyes and its emergence during retinal development," *Vision Research*

12 (1989): 1711–1717. A. T. Hewitt and R. Adler, "The retinal pigment epithelium and interphotoreceptor matrix: Structure and specialized functions," In S. J. Ryan, ed., "Basic Science & Inherited Retinal Disease," *Retina*, 2nd ed., Volume 1, 1994, 58–71.

14. D. Bok, "Retinal photoreceptor disc shedding and pigment epithelium phagocytosis." In S. J. Ryan, ed., "Basic Science & Inherited Retinal Disease," *Retina*, 2nd ed., Volume 1, 1994, 58–71.
15. Ibid.
16. S. M. Raymond and I. J. Jackson, "The retinal pigment epithelium is required for the maintenance of the mouse neural retina," *Current Biology* 5 (1995)" 1286–1295.
17. Hewitt and Adler, "The retinal pigment epithelium and interphotoreceptor matris, 67.
18. Williams, *Natural Selection*, 73.
19. Ibid.

CHAPTER 10
THE MYSTERY OF THE YEAST GENOME

1. We now possess the complete DNA sequences for the genome-libraries of members from all three of life's super-kingdoms: one eucaryote, yeast; one ancient bacteria (archaebacteria), Methanococcus jannaschii; and four ordinary bacteria (eubacteria); Haemophllus influenzae, Mycopiasma pneumoniae, Escherichia coli, and Synechocystis PCC6803.
2. When they did a computer analysis, they found about 960 orthologue groups (remember each group can have more than one gene). Among these 960 groups, they found that only 53 contained representatives from each organism included in the analysis. They found only 80 orthologue groups shared between arcbaebacteria and eucaryotes, but not eubacteria. They found 504 groups shared between the different species of eubacteria but lacking representatives among eucaryotes and archaebacteria.
3. Rebecca Clayton, Owen White, Karen A. Ketchum and J. Craig Ventner, "The first genome from the third domain of life," *Nature* 387 (29 May 1997): 459–462.

Chapter 11
Science and Design

1. Dawkins, *The Blind Watchmaker*, 1.
2. Francis Crick, *What Mad Pursuit: A Personal View of Scientific Discovery* (New York: Basic Books, 1990), 138.
3. Eliot Marshall, "Medline Searches Turn Up Cases of Suspected Plagiarism," *Science* 279 (1998): 473–474.
4. Darwin, *The Origin of Species*, 394 in the Bantam Classic Edition.
5. For a full account, see my book, William A. Dembski, *The Design Inference: Eliminating Chance Through Small Probabilities* (New York: Cambridge University Press, 1998).
6. Ibid.
7. Ludwig Wittgenstein, *Culture and Value* (Chicago: University of Chicago Press, 1984).
8. Bruce Alberts, *Cell* (February 8, 1998).
9. Behe, *Darwin's Black Box*, 39.
10. Dawkins, *The Blind Watchmaker*.
11. John Bodnar, Jeffrey Killian, Michael Nagle, and Suneid RamChandanig, "Deciphering the Language of the Genome," *Journal of Theoretical Biology* 189 (1997): 183.
12. Martin Heidegger, *Being and Time: A Translation of Sein and Zeit* (New York: Harper and Row, 1997), 51.

Conclusion

1. Source obtained from the Internet: Alexander Pope, *Essay on Criticism*, Part ii, Line 15, 1711. Quoted on website www.age-of-the-sage.org/poets/alexander_pope.html

Recommended Bibliography

We are delighted that recent years have witnessed a sharp rise in both the quantity and quality of books and videotapes that explain the problems of Darwinism and the evidence and inferences that point to intelligent design. All of the following books and videos are *highly recommended.* These should be among the first to be added to your library.

At the outset, let us point out a few items of special and strategic importance. First, every serious student of this subject must obtain two resources: Dr. Richard Swenson's extraordinary book *More Than Meets the Eye* and the superb new video *Unlocking the Mystery of Life*, produced by Illustra Media (see our description below). In the "key videos" category, we recommend especially (1) Fred Heeren's *Evidence for God*, (2) Tom Woodward's interview with Michael Behe, titled *Opening Darwin's Black Box*, and (3) *The Chemist's Story*, in which Charles Thaxton shares his amazing life's story as a scientist who stumbled upon the "fingerprint of intelligence" contained in DNA. If you're purchasing key books on Darwinism and Design, it's a good idea to begin with those by Michael Behe, Michael Denton, and Phillip Johnson—and don't forget *The Icons of Evolution* by Jonathan Wells. Amazingly, a unique adult cartoon book, *What's Darwin Got to Do With It?* by Wiester and Newman, may prove to be this list's most powerful introduction to the ideas and evidences of Design.

Those who want to go to the "next level"—into truly challenging material—should start with the three books by William Dembski. Dembski also edited two high-octane collections: *Signs of*

Intelligence (a fabulous new book, notable for its intellectually exhilarating and inspiring chapter by Johnson) and the bulging volume *Mere Creation.* This latter book is a heavy-hitting collection of essays from twenty scientific pioneers—mostly university professors—who came together in 1996, with one hundred fifty others, to discuss Darwinism and intelligent design at a conference in Los Angeles. This conference officially "launched" the Design Movement, so that a reading of this book is like attending that historic conference! If you want an entire "college crash course" on Darwinism's difficulties, get Walter ReMine's meticulous surgical analysis *The Biotic Message.* Final note: All of these resources, like most of those listed below, are available from the three websites that we recommend: apologetics.org, ARN.org, and crsc.org.

Books

Behe, Michael J. *Darwin's Black Box: The Biochemical Challenge to Evolution.* New York: The Free Press, 1996. By far the strongest blow to Darwinism yet, this is the book that "uses a mousetrap to challenge evolutionary theory." Behe demonstrates in easy-to-read language that the biochemical world comprises an arsenal of chemical machines, made up of finely calibrated, interdependent parts—and that this "irreducible complexity" could not have been produced by gradualistic Darwinism.

Buell, Jon, and Virginia Hearn, Eds. *Darwinism: Science or Philosophy?* Richardson, TX: Foundation for Thought and Ethics, 1994. This is the high-powered collection of ten papers given at the SMU Darwinism Symposium, which brought five Darwinists and five Intelligent Design scholars together to debate the central thesis of Johnson's *Darwin on Trial.* Johnson's and Behe's essays are the highlights of this rather technical volume.

Davis, P. William, and Dean Kenyon. *Of Pandas and People* (2nd ed). Dallas: Haughton Publishers, 1994. This is the text developed for use in public schools as a supplement to biology texts on the topic of evolution. Presents the case for intelligent design in a very calm, careful, scientifically scrupulous way, so as to conform to Supreme Court guidelines. A must-read for creationist students.

Dembski, William. *The Design Inference.* Cambridge: Cambridge University Press, 1998. This is the first and the *most highly technical* of Dr. Dembski's three books that lay out the mathematical, statistical, and logical basis for the *detection of design* in the systems, objects, or phenomena in the universe. (Caution: The heart of the book is so highly technical, after chapter one, that your attempt to read it may produce more mirth and head shaking than actual understanding.)

———. *Intelligent Design: The Bridge from Science to Theology.* Downers Grove, IL: InterVarsity Press, 1999. This is Dembski's effort to place his critique of naturalistic creation, and his own "explanatory filter," into more of a layman's level of understanding. This volume is valuable not only for an understanding of Dembski's system for the "detection of design," but also for his explanations of the nature of information and his critique of the worldviews and intellectual fashions, which dismiss *a priori* the possibility of miracles.

———. *No Free Lunch.* New York: Rowman and Littlefield, 2002. Martin Poenie, a Design scholar who teaches zoology at the University of Texas, said this about *No Free Lunch:* "In this book, William Dembski takes his statistical work on inferring design and translates it into an information-theoretic apparatus relevant to understanding biological fitness. In doing so, he has brought his argument for intelligent design into a domain that overlaps current work in evolutionary biology. As I see it, this is a landmark for intelligent design theory because, for the first time, it makes it possible to objectively evaluate the claims of evolutionary biology and intelligent design on common ground."

Dembski, William, ed. *Mere Creation.* Downers Grove, IL: InterVarsity Press, 1998. The "Mere Creation" conference, held in Los Angeles in 1996, featured eighteen plenary talks. Those talks reappear here, with two added, and the twenty contributions comprise one of the most important collections ever to be fashioned by the Design Movement. Since Dembski's "explanatory filter" somewhat dominated the conference, this book is an explanation of that idea and its relevance in relation to other fields.

Dembski, William and James Kushiner, eds. *Signs of Intelligence.* Grand Rapids, MI: Brazos Press/Baker Books, 2001. This collection of fourteen essays first appeared as a special issue of *Touchstone Magazine.* It features contributions by all the leading members of the Design Movement. The essay from Phillip Johnson ("The Intelligent Design Movement: Challenging the Modernist Monopoly on Science") is, in our opinion, one of the most powerful essays he has written and, indeed, one of the most important essays written in the modern history of science.

Denton, Michael. *Evolution: A Theory in Crisis.* Bethesda, MD: Adler & Adler (through Woodbine House), 1986. Denton's work (with the possible exception of Phil Johnson's) is the most powerfully sophisticated and complete critique of the scientific foundations of Darwinian evolution. Very readable and yet massive in its coverage of evidence. It towers over most other books in the field. Order by calling 800-843-7323.

Heeren, Fred. *Show Me God.* Wheeling, IL: Day Star Publications, 2000. This book, revised and updated by the author every four years, is an extraordinarily powerful presentation of key evidences from astronomy, chemistry, physics, and molecular biology, all of which point to a "designed universe." Rarely is a scientific book so educational and yet entertaining at the same time. Heeren's own position on creation is that God appears to have created the universe using the big bang as His mechanism of choice. A few parts of the book may be controversial due to this perspective. Most of the book is a dazzling tour of the evidence for the "fine-tuned" nature of the universe, which has baffled physicists in recent decades.

Johnson, Phillip. *Darwin on Trial* (revised edition). Downers Grove, IL: InterVarsity Press, 1993. Johnson's first book on Darwinism, one of the most important critical manifestos ever written in the past one hundred years, endures as the most important overview of Darwinism in print (besides Denton's work). It surveys many topics, such as current theory, mutations, natural selection, the fossil record, the molecular evidence, the origin of life, as well as philosophical, religious, and educational issues. Johnson concludes that Darwinism is grounded on a powerful philosophical notion called "naturalism," not on compelling evidence. It is written roughly at the level of a college student who is a non-scientist. Perhaps the best part of the book is the Epilogue, in which Johnson surveys what evolutionists have said about his critique and then replies in detail to each of those criticisms.

———. *Defeating Darwinism by Opening Minds.* Downers Grove, IL: InterVarsity Press, 1997. Johnson's "baloney-detector" book. He wrote this specifically for the bright high school student and adults with little science background, to show Darwinists' confused thinking about their own theory. It explains how they try to defend it, using clever rhetorical techniques and by diverting attention away from their problems. This book presents virtually all new material, including new problems with evolutionary theory, not covered in previous books of Johnson.

———. *Evolution as Dogma.* Richardson, TX: Foundation for Thought and Ethics, 1991. This is a delightful booklet that contains a very abridged (seventeen-page) summary of the essence of Johnson's argument in *Darwin on Trial* (minus the discussions of scientific evidence), followed by five responses from a variety of scholars (two are feisty atheistic evolutionists), followed by Johnson's pithy reply.

———. *Objections Sustained.* Downers Grove, IL: InterVarsity Press, 1998. This is a collection of the "best of Johnson" published in scattered magazines and journals, including nine essays on Darwinism and thirteen more on other topics, such as law, cultural trends, and current "hot ideas" making their debut in recently published books.

———. *Reason in the Balance.* Downers Grove, IL: InterVarsity Press, 1995. This is the "sequel" to *Darwin on Trial,* and it surveys the impact of the naturalistic worldview in several scientific areas (astronomy and physics as well as biology), and then traces the role of naturalism in law, ethics, public policy, education, and other areas of the public square.

———. *The Wedge of Truth.* Downers Grove, IL: InterVarsity Press, 2000. This book continues to break new ground as it reports on recent crises and controversies arising in the world of evolutionary biology, such as the Kansas educational brouhaha and the mystery of the source of genetic information. It also explains the "wedge strategy" of the Design Movement and shows that there remains a solid foundation for "reason" to be built on in the new century.

Moreland, J. P. *Christianity and the Nature of Science.* Grand Rapids, MI: Baker Book House, 1989. This is the best survey in print, albeit a bit advanced in its reading level, of the interface between Christianity and the philosophy of science. The opening chapter, which treats difficulties in the "definition of science" (part of the issue of demarcation), is must reading alone.

Newman, Robert and John Wiester. *What's Darwin Got to Do With It?* Downers Grove, IL: InterVarsity Press, 2000. This adult cartoon book, all 144 pages of it, is designed to educate in a clever, imaginative way about the basics of the Darwin vs. Design controversy. To read it is pure fun, but you'll learn more than you might in ten science lectures. Take our word on this.

Pearcey, Nancy and Charles Thaxton. *The Soul of Science.* Wheaton, IL: Good New Publishers/Crossway Books, 1995. This book surveys the development of science and its historic and present relationship to Christianity and reintroduces believers to their rich intellectual heritage. It shows that Christianity, far from being inimical to science, helped to bring modern science into existence. Current debates (including relativity, quantum mechanics, and molecular biology) are treated in their historical, philosophical, and theological relationships.

ReMine, Walter. *The Biotic Message.* St. Paul, MN: St. Paul Science, 1996. This extraordinary book, beautifully bound, is hard to describe. It is simultaneously so massive and meticulously researched, and yet it is also so clearly and powerfully written. It is, arguably, the best all-round analysis and refutation of Darwinism in print. ReMine's thesis is simple but powerful: The evidence of biology delivers a single overriding message. This "biotic message" is that life is the product of a single designer, and the patterns of evidence thwart any attempt to explain it as the result of evolutionary development. The "blurbs" that line up in praise of this book are beyond enthusiastic and come from the likes of Denton, Johnson, and Behe. This book may overwhelm, but it will not disappoint.

Swenson, Richard. *More Than Meets the Eye.* Colorado Springs, CO: NavPress, 2000. This extraordinary book takes the reader on an eye-popping scientific tour of the cell, the brain and its sensory systems, the coiled DNA molecule, and key organs with their sophisticated design. Dr. Swenson even descends down to the level of the "supertiny"—subatomic particles—and up to the vastest reaches of time and space that has been opened up in the scientist's probing of the entire universe. The Creator's awesome greatness is displayed at all levels, but the result is not mere academic knowledge, but a more profound trust in the one who is "very near" as well as "very great."

Thaxton, Charles, Walter Bradley, and Roger Olsen. *The Mystery of Life's Origin.* Dallas, TX: Lewis and Stanley, reprinted 1992. This was the pioneering skeptical overview of chemical evolution that shook up the scientific community when it first appeared in the mid-1980s. It remains one of the most important books on chemical evolution ever published.

Videos

A Chemist's Story (video). C. S. Lewis Society and Trinity College. This video contains a brief televised introduction by Tom Woodward and Arthlene Rippy of the impact of Charles Thaxton upon the world of science. The bulk of this one-hour video is Thaxton's incredible talk at Trinity College, recounting the twists and turns of his amazing discoveries and challenges in the world of science as he pioneered the notion of science's detecting "intelligent design" in DNA. This video touches both the heart and the mind.

Evidence for God (video). Day Star Publications. This half-hour video is like a fast-moving TV documentary from the Discovery Channel. It combines interviews with leading scientists who explain the multiple discoveries that point to a designed universe. Heeren combines several "on the street" interviews, which add a light, amusing touch. Amazingly entertaining, yet also educationally powerful—a "mandatory watching" video!

Icons of Evolution (video). Coldwater Media has produced an extraordinary fifty-minute new video designed to show several of the many flaws in evolutionary theory. It also presents a case for teaching "both sides of the controversy" in public schools and universities. The plight of Burlington, Washington biology teacher Roger DeHart, who was told he could not share any of the scientific problems of Darwinism with his class, is told compellingly. Also, several of Jonathan Wells's "icons of evolution" from his book are beautifully displayed and explained. There is even an excellent section on the Cambrian discoveries in China under Dr. Jun-Yuan Chen, along with the U.S. Congress's recent education law, which directed local schools to teach all sides of the evolution controversy. Order at www.coldwatermedia.com or by calling (800) 889-8670.

Opening Darwin's Black Box (video). C. S. Lewis Society and Trinity College. This twenty-nine-minute video features Tom Woodward's interview with Michael Behe in his DNA lab at Lehigh University. Behe explains, clearly and powerfully, what the main ideas of his book are and how he entered the debate several years ago as he grew more and more skeptical.

The Princeton Chronicles (three-video set). C. S. Lewis Society and Trinity College. In Parts I and II (each thirty minutes) five Princeton University professors explain why they believe in God and have entrusted their lives to Christ; they also show how the evidence from science, from history, and from their personal experience has led them into greater and greater confidence in, and service to, the Creator. Part III features a lecture by Dr. Ed Yamauchi, professor of history at Miami

University of Ohio. It was originally sponsored at Yale and Princeton by Trinity College and details the historical evidence for the resurrection of Christ.

Unlocking the Mystery of Life (video). Illustra Media. This is a spectacularly well-done overview of the rise of Intelligent Design. It features interview clips with Phillip Johnson, Michael Behe, and other key scholars in the movement, with the emphasis on the rising generation of young minds entering the field. All the main ideas of Design, including Behe's "irreducible complexity" and Dembski's ideas about detecting design, are clearly explained. Computer animation, showing the molecular complexity in the cell, is superb and breathtaking. The quality is as good as (or greater than) that of Discovery Channel or PBS documentary. This video, in our opinion, is the single most important resource of any on this list.